Digital Poetics

Studies in the Thought of Paul Ricoeur

Series Editors

Greg S. Johnson, University of Providence, and Dan R. Stiver, Hardin-Simmons University

Studies in the Thought of Paul Ricoeur, a series in conjunction with the Society for Ricoeur Studies, aims to generate research on Ricoeur, about whom interest is rapidly growing both nationally (United States and Canada) and internationally. Broadly construed, the series has three interrelated themes. First, we develop the historical connections to and in Ricoeur's thought. Second, we extend Ricoeur's dialogue with contemporary thinkers representing a variety of disciplines. Third, we utilize Ricoeur to address future prospects in philosophy and other fields that respond to emerging issues of importance. The series approaches these themes from the belief that Ricoeur's thought is not just suited to theoretical exchanges but can and does matter for how we actually engage in the many dimensions that constitute lived existence.

Recent Titles in the Series

Digital Poetics: A Ricoeurian Approach to Digital Technologies, by Fernando Nascimento

From Vulnerability to Promise: Perspectives on Ricœur from Women Philosophers, edited by Sophie-Jan Arrien and Beatriz Contreras Tasso

Paul Ricoeur and Environmental Philosophy, by David Utsler

Refiguring the Sacred: Conversations with Paul Ricoeur, edited by Joseph A. Edelheit, James Moore, and Mark I. Wallace

Paul Ricœur, Philosophical Hermeneutics, and the Question of Revelation, edited by Christina M. Gschwandtner

Reading Ricoeur through Law, edited by Marc de Leeuw, George H. Taylor, and Eileen Brennan

Paul Ricoeur's Renewal of Philosophical Anthropology: Vulnerability, Capability, Justice, by Marc de Leeuw

A Hermeneutics of Contemplative Silence: Paul Ricoeur, Edith Stein, and the Heart of Meaning, by Michele Kueter Petersen

The Phenomenology of Revelation in Heidegger, Marion, and Ricoeur, by Adam J. Graves

Digital Poetics

A Ricoeurian Approach to Digital Technologies

Fernando Nascimento

BLOOMSBURY ACADEMIC
NEW YORK • LONDON • OXFORD • NEW DELHI • SYDNEY

BLOOMSBURY ACADEMIC
Bloomsbury Publishing Inc, 1359 Broadway, New York, NY 10018, USA
Bloomsbury Publishing Plc, 50 Bedford Square, London, WC1B 3DP, UK
Bloomsbury Publishing Ireland, 29 Earlsfort Terrace, Dublin 2, D02 AY28, Ireland

First published in the United States of America 2025

Cover image:
© Thinkhubstudio / iStock / Getty Images

Library of Congress Cataloging-in-Publication Data available

ISBN: HB: 978-1-6669-4223-1
ePDF: 979-8-7651-5437-3
eBook: 978-1-6669-4224-8

Typeset by Deanta Global Publishing Services, Chennai, India

For product safety related questions contact productsafety@bloomsbury.com.

To George Taylor

Contents

Acknowledgments

Let me start by expressing my gratitude to Digital and Computational Studies colleagues: Crystal Hall, Eric Chown, and Mohammad Irfan. It's a privilege to share a supportive and thought-provoking academic environment that has shaped many of the ideas in this book. I would also like to thank my colleagues from the Computing Ethics Narratives and AI Ethics projects—Allison Cooper, Stacy Doore, and Michael Franz—for their support and the many interdisciplinary discussions that fed this book with relevant and fruitful questions.

I am immensely grateful for the generosity of many colleagues who read and offered suggestions on early versions of this work, particularly George Taylor, Lisa Flanagan, and the editors of the Studies in the Thought of Paul Ricoeur series, Dan Stiver and Greg Johnson. I would also like to acknowledge the diligent and competent support of my research assistants, Anya Workman and Ashley Morash.

I would like to thank Candice Hoke for her invaluable advice that significantly enhanced the writing process of this book, leading to extremely productive days of solitude at the Bowdoin Coastal Studies Center.

Finally, I acknowledge the incentive and support of Vanessa, Mateus, Filipe, and Beatriz, my beloved family.

Introduction

Since the dawn of the internet and mobile technologies a few decades ago, digital technologies have been transforming billions of lives across the globe. With a few taps of the screen, we install new applications that will change how we spend our time, consume news, shop, meet loving partners, and make financial investments. The now giant software application ecosystem has grown exponentially in the last decade, driven primarily by the commercial success of some applications and the advertising frameworks that accompany them. During these initial years of the big bang of digital applications, the fundamental criteria for building "good" applications have been technical. Almost everyone involved in technology start-ups focused on product development, and their core objectives were making the software fast, easy to use, and scalable. At best, a certain naivete existed regarding the importance that these applications would have in the lives of people and societies.

Only recently have concerns regarding the ethical and political impacts of such applications become a focus of public debate. Issues ranging from the addiction that some applications create for users, violations of personal privacy, digital breaches of political infrastructure, and the malevolent intensification of polarization in societies have started to gain headlines in major news venues and drawn the attention of concerned citizens and legislators. It has become clear that applications cannot be designed on technical grounds alone and that ethical considerations are fundamental to any technology that so significantly affects the lives of individuals and communities.

As the substantiality and magnitude of the problem become clearer, viable solutions to the new ethical dilemmas generated by digital technologies are still insufficient. The speed with which millions of people integrate new applications into their lives makes ethical and legal frameworks, based exclusively on general rules and laws, quickly obsolete. The complexity of implementing technologies makes external regulatory solutions very fragile, if not completely useless. The rapid distribution of digital technologies makes after-the-fact legal decisions inadequate to prevent new digital artifacts from disrupting social systems and irreversibly impacting individuals and institutions. This scenario, which looks like an ethical perfect storm, sets the background for this book by demanding new ways of thinking about moral problems related to the development and integration of digital technologies into the fabric of everyday lives.

The ecosystem of digital technologies involves a diverse group of social actors who can be broadly categorized into three main types: software developers who master computational techniques and design application and implementation strategies, institutional representatives responsible for product and process regulations, and users themselves who interact with the applications. I take these digital actors as a guiding dimension of my analysis because they allow me to break down my inquiry into spheres of deliberation. Such spheres provide a productive analytical framework for considering the intricate ways in which ethical decisions take place in different phases of the life cycles of digital technologies.

This multidimensional deliberative context leads us to the book's central question: How may we effectively integrate a process of ethical deliberation into the development, use, and regulation of digital technologies? Throughout the book, I address this question by discussing and applying concepts from philosopher Paul Ricoeur's narrative theory and ethics. First, I adopt the Ricoeurian reinterpretation of the concepts of *praxis* (ethics) and *poiesis* (technique) that integrates and intertwines them, providing a framework for considering the incorporation of ethical and technical aspects in developing new technologies. Second, I explore the potential for narratives to be "ethical laboratories" responding to the uncertainty of the unintended consequences of new digital technologies. Third, I apply Ricoeur's "Little Ethics," which integrates abstract moral principles with deliberation in context through practical wisdom to reconcile recent initiatives in the philosophy of technology exclusively based on general principles with the dynamics and unpredictability of new technologies. Fourth, I integrate Ricoeur's emphasis on creative imagination into ethical deliberation and show how software plasticity makes such integration particularly fruitful and effective.

The first chapter introduces the core concepts of emerging digital technologies such as mobile devices, Internet of Things, and artificial intelligence and points to some of their ethical and social implications. Digital technologies possess a set of attributes that together create unprecedented transformative potential: ubiquity, intimacy, and plasticity.

Unlike other technologies of the past, digital technologies are virtually everywhere, and this trend is likely to expand even further as more investments are made in systems based on artificial intelligence and the infrastructure of mobile devices. Being almost ubiquitous, digital technologies mediate actions without the spatial restrictions of other technologies from past decades. Furthermore, the use of individual artifacts is not shared by groups, unlike other hugely impactful technologies of past decades, such as cars and televisions. Unlike most other technological artifacts, digital technologies can metamorphize to each user's tastes and needs. They are personalized and take advantage of this personalization to create usage profiles that can make their adoption even more captivating and addictive.

Emerging digital technologies are also easily extensible. When one buys a microwave, this technology certainly impacts one's cooking habits. However, the microwave's functionalities are relatively stable. In contrast, mobile devices become expandable digital Swiss Army knives. When one downloads a new application, new functionalities become available to smartphone users almost instantly. Consider, for example, food delivery and grocery shopping apps that reshaped the food market during the pandemic and that millions effortlessly installed from the comfort of their home offices and living rooms. Taken together, these features make mobile devices and their software applications agents of unprecedented individual and social change, where technology and ethics are inextricably intertwined. The main objective of the first chapter is to invite the reader to look at what is particular about digital technologies concerning the personal, interpersonal, and social consequences of their widespread deployment in contemporary societies.

The peculiarities of digital artifacts lead us to how these technologies' inception, design, and implementation can integrate ethical and technical issues. The second chapter lays out the landscape of the dichotomy between technique and ethics in the last decades of software development up to the current debate, which often focuses on high-level principles-based codes of ethics. Following this contextual analysis, I explore and expand the inquiry into software engineering practice based on Ricoeur's theory of action, particularly through the concept of practice that connects his theory of action to his "Little Ethics" and underpins our reflection on the intrinsic ethical nature of all technical actions within the practice of software development. Such an approach brings an additional dimension of meaning to the forefront in the development of digital technologies in which appraisals of an evaluative (and subsequently normative) nature are attached to precepts of doing (something) well.

Once the intrinsic integration between technique and ethics in digital technologies is posed, we must ask ourselves about the best ethical approach so that such integration maximizes the possibility of developing or enhancing digital technologies to foster the common good. The third chapter addresses the challenge of integrating ethics into new technology development by introducing the concept of narrative deliberation. Narrative deliberation articulates Ricoeur's narrative theory and "Little Ethics," focusing on the specific challenges of ethical decisions related to digital technologies. In the first section, I explain Ricoeur's concept of critical practical wisdom, highlighting three characteristics that are particularly suited to address the ethical dilemmas that digital technologies exponentialize. First, critical practical wisdom puts deontological (obligation-oriented) and teleological (ends-oriented) principles in a dialectical tension, creating a rich deliberative framework where general rules and particular cases challenge and complement each other. Second, through ethical imagination, critical practical wisdom dissolves simplifying and generalizing polarizations and transforms them into continuums in which the peculiarities of each decision-making instance can be better evaluated. Third, critical practical wisdom recognizes the need for a multiple-person, interdisciplinary deliberative body that Ricoeur suggests resembles public debate. Each feature will play an essential role in analyzing the ethical dilemmas of digital technologies that subsequent chapters explore.

In the subsequent sections of chapter 3, I then propose expanding Ricoeur's concept of critical practical wisdom based on his narrative theory. At the heart of the concept of practical wisdom is deliberation. I suggest that Ricoeur's concept of threefold mimesis provides a privileged analytical framework for rethinking deliberation. First, it allows us to see the decision not just as a sentence that wraps up a juridical process but also as the configuration (mimesis 2) of a prospective narrative through creative imagination that intends to re-establish the concordant discordance put in jeopardy by the ethical issue in question. Second, it frames deliberation as a plurivocal process that represents both what happened and what shall happen as a configured plot integrating various arguments and perspectives. Third, it highlights the ethical relevance of refigurations (mimesis 3) triggered by historical and fictional narratives to develop an ethical sensibility that is critical for deliberating well about technological artifacts whose contexts of use are complex and uncertain.

In the final three chapters, I apply, probe, and refine the concept of narrative deliberation through different angles of analysis based on spheres of deliberation defined, respectively, by developers, users, and institutions. They function as general types to analyze different contexts of applying narrative deliberation in digital technologies. Considering each sphere of deliberation at a time, I apply four essential characteristics of narrative deliberation to each of the contexts defined by these actors: (1) the contextualization of moral principles in the search for decisions that maximize the common good in new situations brought about by digital technologies, (2) the need for public debate for good deliberation, (3) the unique intertwining between technique and ethics in decisions involving software-based entities, and (4) the exercise of the narrative imagination to overcome practical aporias in the deliberative process.

Taken together, these four characteristic dimensions of narrative deliberation address the demand for new ethical approaches in developing digital technologies and represent a poetic response to the uncertainties and ethical aporias of the digital world. Based on Ricoeur's narrative theory and ethics and in dialogue with recent developments in the philosophy of technology and media studies, these dimensions form the axes of what I call *Digital Poetics*.

Digital Poetics is about making digital things, and some conceptual premises help to clarify the meaning we propose for the term "poetics" in this text. The first conceptual premise is that Digital Poetics is about *poiesis* in the classical sense of making meaningful things that change the world. These meaningful things range from everyday objects like a chair or a boat to a poem or a Sophoclean tragedy. It is also in this sense that Ricoeur says that narrative is *poiesis*: it is a creative activity that brings new meanings into the human world.

The second conceptual premise is that the *poiesis* of digital things always has an ethical dimension. Just as narratives describe and prescribe a world of the text, intertwining *poiesis* and *praxis*, Digital Poetics is always an entanglement of ethical values realized in algorithms, communication protocols, user interfaces, hardware, and software. As we will repeat several times throughout the text, we start from the premise that every digital creation is *both poiesis* (technique) and *praxis* (ethics).

The third conceptual premise is that the transformation of meaning brought to the world by Digital Poetics is collaborative. Developers are the direct creators of digital technologies and impose certain contours on digital artifacts, but users co-participate in changing these artifacts by choosing ways to integrate them into their lives and goals. Institutions must also be co-authors of the transformations brought about by Digital Poetics; they must inform how these technologies align with community values. Digital Poetics is a joint process that takes place in various spheres of deliberation and agency that complement and affect each other.

There is admittedly a utopian dimension to the Digital Poetics project. It is utopian in the sense of breaking with the status quo in which technical development is seen as completely separate from its ethical implications. The recent advances and proliferation of tools in artificial intelligence, social networks, and virtual reality have increased public debate about digital technologies' social and personal impacts. However, the current research and development structures of digital technologies, which are primarily driven by capital accumulation, make the proposal of Digital Poetics both a utopia of creating a possible world and an ethical imperative.

Taking the utopian ideal to the limit, Digital Poetics proposes envisioning digital technologies as new ways of signifying human experience shaped by ethical intents, analogous to Ricoeur's beautiful description of the poet's activity: "Poetry is more than the art of making poems. It is *poiesis*, or creation in the largest sense of the word. It is in this sense that poetry is equivalent to primordial dwelling; man dwells only when poets exist in the world."[1]

Note

1 Paul Ricoeur, *The Conflict of Interpretations: Essays in Hermeneutics* (Evanston: Northwestern University Press, 1974), 467.

1

A Brief Introduction to Digital Technologies

In this book, we will investigate the ethical aspects of the art of producing digital technologies, their uses, and social impacts. Digital technologies permeate so many dimensions of contemporary life that they almost need no introduction. However, daily interaction with smartphones, mobile apps, computers, and websites renders opaque a series of peculiarities of digital technologies that will be essential to our investigation of Digital Poetics. This chapter offers a brief analysis of the defining characteristics of modern digital technologies, using smartphones as a paradigmatic example. These devices share some key essential characteristics, such as ubiquity, personalization, and adaptability, which we will explore further in the following sections. Taken together, these essential characteristics provide us with a better perspective for understanding the transformative power of digital technologies and the unique ethical challenges posed by their design and implementation, use, and regulation. We begin with a brief look at the material and functional aspects that underlie all digital technologies and differentiate them from other technological artifacts.

Regarding their material composition, there are two central attributes of digital technologies. First, they represent and manipulate information based on a reduced set of electrical signals. In most cases, these electrical signals are translated into binary digits, becoming the basic information representation units in digital technologies. Second, the information is processed by electronic circuits that implement logical operations quickly and consistently. From a functional point of view, digital technologies are typically related to encoding, storing, transmitting, and processing information, which is why they are often referred to as "Information and Communication Technologies (ICTs)."[1]

In recent decades, there has been an explosion of new applications based on this small set of signals and logical operations. The speed with which digital technologies have advanced in functionality, social relevance, personal impact, and agency in recent decades is impressive.[2] These advances constantly surprise technologists, and recent developments in the ability to automatically generate texts and images through techniques called generative artificial intelligence (AI) have surprised even researchers and scholars involved in cutting-edge research in this field.[3]

It is, therefore, not unexpected that philosophical-ethical reflection on the impacts of technology has only recently been dealing more extensively with the distinctive characteristics and peculiar impacts of digital technologies. It is also understandable that Paul Ricoeur, who wrote most of his work in the twentieth century,[4] has not explicitly thematized the ethical impact of digital technologies.[5] However, as we will show throughout the book, Ricoeur did discuss at various points technology's ethical and social impacts. Furthermore, several concepts from his philosophy provide invaluable frameworks for a deeper understanding of the ethical impacts of digital technologies.[6] One of the few times Ricoeur explicitly discussed the ethical aspects of digital technologies was at a roundtable on practical wisdom in 1997. During the question and answer session, Ricoeur was asked about the impact and risks of using computers to make ethical and moral judgments in the legal context. Ricoeur answered the question by saying, "There's one thing we must always remember, and that's that computers don't think. There are people who use computers. Computers are instruments of communication, but they don't invent messages or think."[7] Ricoeur argued that computers were only intermediaries in the storage and transmission of information, but they didn't essentially affect human agency in ethical deliberations: "You can't change the fundamental situation of people judging people."[8]

It is impossible to know how Ricoeur would answer the same question today when artificial intelligence-based programs are used in various jurisdictions in the United States to decide the risk of recidivism of defendants requesting parole.[9] Or when large language models create summaries of scientific articles[10] and write coherent essays, compose letters, and automatically reply to electronic messages. Digital technologies' functionalities, amplified by recent artificial intelligence algorithms, have been transforming the possibilities of human agency in ways that were extremely hard to predict a few decades ago.

The recent advances in digital technologies demand a new philosophical inquiry[11] on the impact of these technologies on ethical deliberation. Digital technologies do not provide simply inert extensions of the relationship between doctors and patients, judges and defendants, writers and readers, or, in short, between any agents and recipients of ethical actions. As we will investigate throughout the book, digital technologies transform human agency in unprecedented ways and increasingly demand ethical engagement to deliberate their repercussions in the personal, interpersonal, and institutional spheres of human experience. This chapter focuses primarily on describing some functional characteristics of digital technologies, laying the groundwork for a later examination of their ethical implications, which will be the focus of the succeeding chapters. The following sections of this chapter analyze other peculiarities of digital technologies that differentiate them from other technologies and increase their impact on human agency and societal institutions.

Digital Plasticity

Representational Flexibility

Digital technologies model different forms of physical processes and experiences. The simplicity of the basic representational and computational units of digital technologies allows them to represent and often replace various natural and social phenomena and processes, creating sophisticated levels of abstraction and representation of information on the basis of the intentional combination of basic-level binary digits and logical operators.[12] Consider, for instance, a movie streaming application. It translates acoustic signals into sequences of binary digits processed by logic operators that, combined with other hardware components such as the microphone, produce the movie sound. The exact same representational and processing basic units also translate visual images and display the image frames of the movie. Furthermore, the instructions of the application itself are encoded using binary digits and basic logic operators. This representational flexibility is one of the fundamental characteristics of digital technologies.[13] It allows digital applications to model completely different high-level processes and for that information to be transmitted and processed by the same digital machines using similar computational mechanisms. Representational flexibility also allows digital artifacts to adjust their behavior and appearance according to their users, creating the possibility of personalized experiences.

Personalization

Personalization is a distinctive feature of digital technologies. Each digital artifact reflects the preferences and tastes of its owner. For example, users can customize the screens of laptops, smartphones, and tablets with a myriad of wallpapers, layouts, and widgets. This customization capability extends to applications, such as social networking ones, where users can create detailed profiles with photos, aliases, preferred colors, and font styles. Digital personalization provides a tailored user experience that makes each device a unique reflection of its owner.

At the same time, this individualization is accentuated by recommendation systems,[14] which learn and adapt to user behavior patterns. These systems are embedded into various applications, from video streaming platforms[15] to online shopping services.[16] They analyze past interactions to suggest products, posts, movies, and music that align with the user's interests. For example, video streaming services[17] present a personalized homepage based on the user's viewing history and preferences derived from affinity groups that emerge from correlated statistical regularities between various users of a digital artifact. This targeted approach captures user attention more effectively, potentially leading to addictive behaviors that can severely impact user agency.[18] The chameleon-like nature of emerging technologies is a fundamental facet of "digital materiality." These technologies can create a detailed model of users' desires and preferences, profoundly affecting attention and agency.[19] As devices become more tailored to individual interests, they become more engaging and impact more practical aspects of daily life.[20] This phenomenon raises important questions about the extent and intensity of these devices' influence on users, especially considering the possibility of their shaping or even determining choices and behaviors.

Digital personalization makes this type of technology extremely convenient, and such convenience impacts the deliberative processes of users who feel more inclined to carry out actions mediated by digital technologies. Through AI-based recommendation systems, digital technologies "remove practical hurdles, almost immediately fulfilling our desires. For example, e-commerce platforms are built to facilitate frictionless consumption. Choosing the best gift can be done from the couch using a friendly website interface."[21] The adaptability of the behavior of digital technologies also makes them convenient for predicting with some accuracy the choices that users would have made without the mediation of these technologies. "Users realize with delight that a product recommended to them is precisely what they wanted—without them even knowing it."[22]

Thus, the magnitude and nuanced level of personalization of AI-based digital technologies represent a significant difference from non-digital technologies that usually leave little room for adjustment according to user preferences and have little agency to metaphorize themselves according to the tastes of their "owners." Recommendation systems based on sophisticated artificial intelligence algorithms can devise predictive models based on the behaviors of each user and on the massive databases of similar users' digital footprints, personalizing digital technologies in the image and likeness of their users' desires. While these devices offer valuable opportunities for tailored experiences and convenience, they also present unique challenges and dangers to user autonomy and agency. Understanding these challenges is crucial and points to the need for deliberative processes that align the incredible conveniences of digital personalization with broader goals of lives fulfilled with and for others in sustainable natural environments.

Functional Extensibility

Digital technologies are also distinguished by their functional extensibility. Unlike technologies with predefined functionalities, such as cars, television sets, and ovens of past decades, whose functions did not change significantly over the artifact's life cycle, mobile devices are digital multi-tools that gain new capabilities with each app installation. Food delivery and grocery shopping apps, which became prevalent during the COVID pandemic, are paradigmatic examples of how these devices can metamorphize to meet users' needs in the face of new contexts. Digital devices are not just artifacts; they can also be platforms that allow for creating completely new affordances and impacts on users' agency. Consider, for example, how the functionality of a smartphone is dramatically transformed by installing applications such as Instagram, Uber, Weather, Venmo, and so on. Smartphones, computers, smartwatches, and other digital technologies are *general-purpose technologies*. Daron Acemoglu and Simon Johnson point out that they "provide a platform on which a myriad of applications can be built and potentially generate benefits—but sometimes also costs—for many sectors and groups of people."[23] The same ability to expand functionalities occurs with web-based applications. It is possible to access a multitude of new applications that appear as extensions of the functionalities available on the computer using the same internet infrastructure and the same web browser. This is also the case with video game platforms that provide application stores where new games can be installed and accessed via the same digital gaming console.

Before web-based and mobile applications, digital technologies had restrictions on expanding their functionalities similar to technologies with other materialities. For example, if there was a problem with the software on a mobile phone in the early 2000s, you had to take it to a service provider to update the software. This meant that the development cycle for mobile applications was much longer to prevent serious issues that would cause recalls and profit losses. What's more, when you bought mobile devices in the 1990s, almost the entire set of their functionalities was predefined and remained the same throughout the life of the device because it wasn't possible to install applications remotely. Similarly, before the proliferation of web-based applications, it was necessary to buy software products that had their functionality pre-established, such as the popular text editors and spreadsheets that extended the functions of home computers but had well-defined and relatively fixed functions throughout their life cycle.

Recent web-based and mobile applications allow developers much more flexibility in distributing and updating digital artifacts. Every time one enters an address on a website, a request is sent to a computer (usually called a server) located in the developer's digital infrastructure that returns the content of the page that your browser (usually called a client) displays. Thus, when necessary, developers can update the software and algorithms on the server, and the changes are automatically reflected in all future results on the clients of that digital artifact.

Something similar happens with mobile applications. Mobile operating systems such as iOS and Android allow application versions to be updated remotely, a process called "Over-The-Air" (OTA) update. This possibility dramatically transformed digital technologies, as it allows corrections to be made almost all the time without users even realizing that something has changed in the functionality of their artifact. With OTA updates, Digital Poetics has become much more malleable and iterative. Unlike other ways of "doing," developers of digital technologies have an almost instantaneous ability to impact users' experience throughout the technology usage lifecycle. In chapter 2, we will discuss the implications of this functional extensibility in the development life cycle of digital technologies and software development practice, and in chapter 4, we will discuss the impacts of functional extensibility on developers' ethical deliberation.

Digital Pervasiveness

Ubiquity of Access

The ubiquity of digital technologies, such as mobile phones, has radically transformed how we interact with the world around us and is one of the central characteristics of Digital Poetics that affects human agency and ethical decisions. Unlike previous technologies, many digital technologies can be accessed from virtually anywhere. This universal access capacity is rooted in the expansion of cellular[24] and Wi-Fi networks. Cellular networks allow continuous connectivity on the move, while Wi-Fi networks offer an efficient, high-speed means of connecting digital technologies to the world through the internet indoors. The transition between these networks is generally fluid, providing a seamless connectivity experience that few previous technologies could offer. Furthermore, the ubiquity of digital technologies breaks down the traditional barriers between internal and external spaces. Consider, for instance, two other immensely impactful technologies of recent decades: cars and television. Cars were mainly used for outdoor travel but remained idle in garages or parking lots, rarely affecting how we lived indoors. Televisions, predominantly domestic artifacts, changed home entertainment and family dynamics, but we could rarely interact with them once we left our homes. By contrast, many digital technologies move effortlessly among all these diverse physical environments. They accompany their users in almost every location, from the dinner table to business meetings, parks, commutes, international flights, and beyond.

From the point of view of interpersonal relationships, this ubiquity has resulted in a significant change in how people interact. Previously, communication technologies were confined to specific locations: landlines in homes or offices and phone booths in public places. Now, communication is ubiquitous, enabling constant interactions regardless of physical location.[25] Users of mobile digital devices can stay connected with family, friends, and coworkers regardless of whether they are at home, in the office, or on the move. This ability to communicate constantly has shaped new social norms and expectations regarding availability and rapid response. Richard Ling sums up this impact of digital technologies as dialectic between more connectivity and new expectations of interpersonal relationships: "At one turn, it gives us the freedom to move about while maintaining contact; at the next turn, it ties us securely into the expectations of our social networks."[26]

The social expectations transformed by digital technologies' ubiquity go beyond interactions with family and friends and have reshaped working relationships. With constant access through digital technologies, it is possible to work from almost anywhere, and this brings more flexibility and can be reflected in the benefits of shorter commute times and more flexible schedules. However, several studies, particularly following the effects of the COVID pandemic, point to possible negative effects of the constant expectation of availability to solve work-related problems that interrupt other types of relationships and preclude immersion in restorative activities.[27]

From interpersonal relationships to working expectations, the line between being "online" and "offline" has become increasingly blurred, reflecting the deep integration of digital technology into almost every aspect of the contemporary human experience in more industrialized countries.[28] A recent survey pointed to the growing perception of the pervasiveness and impact of smartphones in the daily lives of Americans: "The percentage of U.S. adults saying they use their smartphone 'too much' has increased markedly in recent years, rising from 39% when Gallup last asked this in 2015 to 58% today."[29] However, in an apparent paradox, the survey also highlights the perception that, in general, the use of smartphones has brought benefits: "two-thirds of Americans say smartphones have made their life better."[30] And overall, the survey reinforces the perception of the impact of smartphones on users' lives: "half of Americans agree with the statement: 'I can't imagine my life without my smartphone.'"[31] Throughout the book, we will suggest that this apparent paradox can be minimized through a process of ethical deliberation that integrates the recognized benefits of digital technologies with the more holistic goals of a fulfilled life.

Instant Global Distribution

Another critical aspect of digital pervasiveness is the creation of a distributed software development ecosystem in which smartphone and computer manufacturers allow external developers to install applications on their digital architectures.[32] This new ecosystem for creating and distributing digital technologies takes advantage of the stability and transferability inherent in digital materiality. The digital development ecosystem has generated two unique characteristics of digital technologies: virtual application stores and a multitude of application development companies. Both phenomena are essential factors in Digital Poetics, as we will explore in chapter 2. Virtual app stores, such as the Apple App Store and Google Play,[33] allow users to search for new functionalities, which have significantly expanded the impact of digital technologies on their lives. App development companies leverage digital platforms to reach their users directly on any device connected to the virtual app stores. "By 2016, the [Apple] App Store had more than two million apps available for download, with about 100,000 new or updated titles arriving every week. Using the App Store was safer than downloading a conventional program from the Internet and offered something the Web lacked: a built-in payment system."[34] Unlike other technologies, digital plasticity allows digital technologies to be distributed almost instantaneously worldwide. A new version of the Uber app can be spread through digital architectures to millions of users overnight and change the processes they use to request rides or receive pay for transportation services, affecting the agency of people and institutions worldwide almost instantaneously.

Web applications accessed from desktop computers and notebooks also have the same capacity for global and instantaneous impact. A recent example is web applications for accessing information from large language models, such as ChatGPT, Bard, and Claude. The figures for digital pervasiveness are impressive. Enabled by digital architectures, ChatGPT, launched in November 2022, reached 100 million users in two months.[35] Such an impressive growth points to the *acceleration* of digital pervasiveness. To obtain an idea of this acceleration, we can look at the adoption rate of other recent technologies that have immense penetration and social impact. The popular social network Instagram, launched in 2010, took two and a half years to reach the 100 million user mark, which already marked a distinctive feature of digital technologies over other types of technology. This pace was accelerated by the other popular short video app, TikTok, which was launched in 2016 and reached the same number of users in just nine months. These figures show not only that the speed of penetration of digital technologies into users' lives is very fast but that this phenomenon is accelerating through the architectural capabilities of almost instantaneous global distribution.

One of the issues that permeates this entire book is the gigantic challenge of the deliberation of new ethical and political dilemmas brought about by the acceleration in the impact of digital technologies. Digital Poetics is not just a set of new techniques and tools for the exponentially faster creation of digital artifacts. Rather, it is an integrated form of conception, deployment, regulation, and use of digital technologies that integrate new capacities for doing (*poiesis*) with ethical actions (*praxis*) to foster human capabilities. A technicist vision that reduces digital technologies to their merely technical aspects is fallacious. As Acemoglu and Johnson's extensive survey of the history of technology suggests, "a thousand years of history and contemporary evidence make one thing abundantly clear: there is nothing automatic about new technologies bringing widespread prosperity. Whether they do or not is an economic, social, and political choice."[36]

Digital plasticity and its pervasiveness in people's lives increase the urgency of ethical discussions that integrate new forms of ethical imagination to address novel effects of these technologies, such as those generated by large language models in educational processes and systems. Given the global penetration of digital technologies, these deliberations also need to integrate the plurality of perspectives of people affected by digital technologies. Throughout this book, we will discuss how a proposal for narrative deliberation that combines ethical imagination and plural deliberation offers support for a new form of ethical deliberation, one that considers the fundamental characteristics of Digital Poetics.

New Forms of Agency Distribution

Digital technologies have transcended their role as mere tools, morphing into active participants in our world. Unlike their more passive predecessors, they don't just respond to our commands; they initiate, adapt, and collaborate in ways that blur the lines between human and machine agency. Consider, for example, the digital thermostat that can adjust itself and learn the usage patterns of the rooms in the house to optimize energy consumption and thermal comfort. Or voice assistants that can answer complex questions and place orders on shopping websites. Or smart refrigerators that notice the milk is running low and add a new entry to the grocery list. The increased impact of digital technologies on human agency is one of the central aspects of the Digital Poetic ethical-philosophical analysis.

Much of the increased autonomy of emerging technologies derives from machine learning algorithms. These algorithms, usually categorized under an umbrella called "artificial intelligence," are different from traditional algorithms because they create their prediction models based on a set of training data and optimization objectives. Coeckelbergh explains this process in relation to the different strategies for creating the rules that make up the heuristic models of digital technologies: "In contrast to expert systems, which rely on human domain experts who explain the rules to programmers who then code these rules, the machine learning algorithm finds rules or patterns that the programmer has not specified. Only the objective or task is given."[37]

The models generated by artificial intelligence are predictive and agentive. Based on the objectives defined by the developers, AI-based digital technologies can define the next post we will see on our social media timelines, determine the risk of skin cancer in an image, and initiate irrigation processes based on optimal soil moisture patterns for a certain crop type. In this sense, there is a more significant share of agency attributed to emerging digital technologies. As we shall discuss further in the following chapters, thinkers such as Bruno Latour propose a model of shared action between various human and nonhuman *actants*.[38] Actants collaborate in the execution of actions. Emerging digital technologies based on predictive models have increased capacities to act and change the balance of humans and nonhuman actants in multiple processes where actants were uniquely human a few years ago.

Non-AI-based digital technologies have already accelerated the shift in the weights of the actants' participation in actions through flexible models that automate steps that were previously carried out by human actants. For example, the defrosting option of various microwaves automates the heat duration and intensity needed to defrost multiple types of food without a human actant having to calculate and adjust the microwave's parameters step by step. However, the models that estimate the durations and intensities based on the type of food and its weight had to be explicitly coded into the microwave software by human programmers who, in discussion with food engineers, created a defrosting model. A possible AI-based microwave could create a defrosting model itself on the basis of past defrosting data patterns and visual recognition of the food, so the human actant will only put the food in the microwave and ask it, using natural language commands, to defrost or cook it.

In addition, digital technologies can collaborate among themselves, creating a collective digital agency that grows with the gradual increase in the ability of digital devices to communicate with each other, forming sophisticated agency networks usually referred to as the "Internet of Things." Luciano Floridi suggests that there is a scale of technological "orders" in which the agency of digital technologies increases to the point where various tasks are carried out by interactions among digital technologies only, without being initiated by human agents.[39] An example of this "order" of agency between digital artifacts is the car communicating with the garage door to open it when it approaches the house. The garage, in turn, communicates with the lights in the house and the thermostat to adjust the environment according to the climate and the preferences of the house's inhabitants.[40] This kind of autonomy, beyond just automation of predefined processes and parameters, is a fundamental characteristic of emerging digital technologies that create new ethical questions for Digital Poetics because they affect human agency in unprecedented and substantial ways.

Refiguration of Meanings

Digital technologies affect human agency even more profoundly through the resignifications of meaning they propose through their screens and interaction interfaces. Thus, they not only affect actions but also how we signify and interpret the actions that come to be mediated by digital technologies. A paradigmatic example of the interpretive mediations embedded in digital technologies is the metaphors developers use to represent digital artifacts' functionalities and interfaces based on non-digital experiences. Through digital metaphors, developers invite users to *see* a touch on the surface of a smartphone *as* selecting an option or to *see* a computer screen *as* a workspace. All the complex technical details of implementing touch detection on the surface of the cell phone screen and file management on the computer are translated to users through the layer of meaning proposed by digital metaphors.[41]

Given digital ubiquity, one of the tasks of Digital Poetics is to make digital technologies accessible and understandable to non-technical users. The complexity of digital technologies, often loaded with intricate implementation details, can be daunting for many users. To bridge this gap, designers employ metaphors to translate these complexities into the semantic horizon of ordinary users. This translation process is not just about simplification; it's about creating a relatable and intuitive understanding of digital functionalities. These digital metaphors are powerful tools for making abstract or complex digital concepts more tangible. For example, conceptualizing a database entry as a "friend" in social media contexts or interpreting a "thumbs up" icon as an expression of approval helps users quickly understand and interact with digital interfaces without understanding the underlying technical mechanisms. These metaphors give users a familiar framework, making digital interactions friendlier and less intimidating.

Because of digital plasticity, digital technologies can realize these metaphors in applications that model various types of actions, translating and transforming them into digital experiences. The limitations imposed by the COVID pandemic, for example, have accelerated this facet of the digital: "people were able to attend birthday parties, weddings, movie nights, book clubs, dance raves, meditation sessions, yoga classes, and religious services, including funerals."[42]

The pervasive use of these digital metaphors and models in our daily interactions has profound implications for non-digital experiences. They gradually reshape our understanding and expectations of actions and expressions in both the digital and physical worlds. For example, expressing appreciation, which once might have involved a range of nuanced behaviors, can be oversimplified by clicking a "like" button. This not only affects how we express emotions in the digital domain but also has the potential to refigure our behaviors and expectations in real-life interactions. The ease and immediacy of digital expressions can refigure how we perceive and value more traditional, often more demanding, forms of communication and interaction. Eric Chown and I have corroborated the importance of these refigurations of meanings, arguing from a cognitive and hermeneutic perspective that digital metaphors affect how we act in the world and how we signify fundamental human experiences such as conversations, friendships, dating, the acquisition of goods, and entertainment.[43] Thus, the initial goal of employing digital metaphors to explain the functionalities of digital technologies evolves into something much more consequential. It becomes a process of redefining meanings and experiences not mediated by digital technologies. Initially designed as bridges to understanding, these metaphors can fundamentally alter human behavior and interactions. They reflect and shape the horizon of meanings of how we communicate, interact, and understand the world around us.

In *Transcoding the Digital: How Metaphors Matter in New Media*, Marianne Van Den Boomen investigates how metaphors are crucial in the use and understanding of digital artifacts.[44] For her, "the practice of digital code exchange can only be articulated, perceived, and conceived when it is translated into metaphors."[45] She has extensively analyzed how metaphors influence digital experiences across different layers, employing a new media studies approach. Her research begins with user interfaces and extends to social discourse, highlighting each level's significant impact on culture and society. This transformative power of digital metaphors highlights the wider implications that these experiences have on society and culture.[46] Digital metaphors reframe and refigure meanings of social phenomena, interpersonal relationships, and personal experiences.

Digital metaphors highlight that digital technologies operate not only on a pragmatic level, changing actions, but also on a hermeneutic level, altering how we interpret and value our experiences in the world. Therefore, the analysis of Digital Poetics must take into account how digital technologies transform both the practical aspects of agency and the horizons of meaning of users and institutions. And, as we shall explore further, the semantic and action theories proposed by Ricoeur provide an excellent conceptual framework for analyzing the hermeneutic dimension of Digital Poetics.

Corollaries for Digital Poetics

The combination of the peculiar characteristics of digital technologies brings about a series of consequences for Digital Poetics that we briefly describe in this final section and employ in subsequent chapters as a backdrop for the discussions of the ethical implications of digital technologies. We have organized these consequences based on *Digital Poetics' spheres of deliberation*, which represent the primary contexts in which ethical decisions about the development and use of digital technologies must be made. In all spheres, Digital Poetics is marked by (1) a multidimensional impact on human agency, (2) the acceleration of the creation and deployment of functional innovations, and (3) the malleability of the design, updating, and use of digital technologies. These traits distinguish digital technologies from other technologies and raise unique ethical questions for the philosophy of technology.

In the deliberative sphere of digital technology developers, instant global distribution creates two accountability pressures. Unlike other technologies whose development and distribution times have been constrained by the complex logistics of physical transportation of raw materials and finished products, the development and distribution of various digital technologies, such as the internet and smartphones, leverage distribution architectures with extreme capillarity and pervasiveness. Thus, the developer sees the impact of *poiesis* (technique) on users almost instantaneously. What's more, as we'll describe in detail when we analyze the history of software development practice in chapter 2, the users of digital technology are multinational, multicultural, and in much greater numbers than other non-digital technologies.

On the other hand, Digital Poetics brings new possibilities for the agency in the developer sphere, which allows for variations in development cycles to accommodate new ethical concerns. With instant global distribution capabilities, developers can make incremental deployments and check the repercussions of their modifications on gradually larger groups of users. In addition, digital architectures provide monitoring and feedback mechanisms that other technologies don't have. Digital Poetics has a bidirectionality between developers and users that allows developers to remotely monitor errors in application behavior or evaluate how certain functionalities are used after an over-the-air deployment.

In the sphere of user deliberation, Digital Poetics is marked by the personalization and functional extensibility of digital technologies that impact almost every dimension of personal agency. Users are constantly faced with decisions about when, how much, and for what purposes they should use or extend their collaboration with digital technologies. For example, students must deliberate on collaborating with large language models to carry out their academic work. These deliberations highlight the intricate relationship between practical and ethical decisions in Digital Poetics in the user sphere.

Finally, in the institutional or public sphere, plasticity and pervasiveness create substantial challenges for deliberators who are faced with technologies whose accelerated innovation, technical complexity, and wide-ranging social impacts challenge the structures and processes of deliberation, regulation, and enforcement of public policies and the definition of ethical principles and guidelines that Digital Poetics should follow and adopt.

Notes

1 Luciano Floridi, *The Fourth Revolution: How the Infosphere Is Reshaping Human Reality* (Oxford: Oxford University Press, 2014), vi.
2 Nicole Aschoff, *The Smartphone Society: Technology, Power, and Resistance in the New Gilded Age* (Boston: Beacon Press, 2020).
3 Will Douglas Heaven, "The Inside Story of How ChatGPT Was Built from the People Who Made It," *MIT Technology Review*, March 3, 2023, https://www.technologyreview.com/2023/03/03/1069311/inside-story-oral-history-how-chatgpt-built-openai/.
4 Ricoeur's last major work, *The Course of Recognition*, was published in 2005.
5 David M. Kaplan, "Paul Ricoeur and the Philosophy of Technology," *Journal of French and Francophone Philosophy* 16, no. 1/2 (January 26, 2006): 42.
6 Ernst Wolff, "Ricoeur's Polysemy of Technology and Its Reception," in *Interpreting Technology: Ricoeur on Questions Concerning Ethics and Philosophy of Technology*, edited by Wessel Reijers, Alberto Romele, and Mark Coeckelbergh (Lanham: Rowman & Littlefield, 2021), 3.
7 Université de Picardie, Jeffrey Andrew Barash, and Mireille Delbraccio, *La Sagesse Pratique: Autour de L'oeuvre de Paul Ricoeur: Colloque International, Université de Picardie Jules Verne, Amiens, 5–7 Mars 1997* (Centre national de documentation pédagogique, 1998), 18–19.
8 Picardie, Barash, and Delbraccio, *La Sagesse Pratique: Autour de L'oeuvre de Paul Ricoeur: Colloque International*, 19.

9 Meredith Broussard, *More than a Glitch: Confronting Race, Gender, and Ability Bias in Tech* (Cambridge, MA: MIT Press, 2023), 54.

10 Holly Else, "Abstracts Written by ChatGPT Fool Scientists," *Nature* 613, no. 7944 (January 2023): 423.

11 See, for instance, Guido Gorgoni, who commented on the same remark by Ricoeur. Guido Gorgoni, "Reflecting on Identity and Autonomy in a Datafied Society with Paul Ricœur," in *The Challenges of Autonomy and Autonomy as a Challenge Thinking Autonomy in Challenging Time* (SVK, 2022), 97–108.

12 Peter J. Denning and Matti Tedre, *Computational Thinking* (Cambridge, MA: MIT Press, 2019), 54–57.

13 Mark Priestley, *A Science of Operations: Machines, Logic and the Invention of Programming* (London: Springer Science & Business Media, 2011), 147–53.

14 Michael Schrage, *Recommendation Engines* (Cambridge, MA: MIT Press, 2020).

15 Carlos A. Gomez-Uribe and Neil Hunt, "The Netflix Recommender System," in *ACM Transactions on Management Information Systems*, 2016, https://doi.org/10.1145/2843948.

16 Brent Smith and Greg Linden, "Two Decades of Recommender Systems at Amazon.com," *IEEE Internet Computing* 21, no. 3 (May 2017): 12–18.

17 Gomez-Uribe and Hunt, "The Netflix Recommender System," 1–19.

18 Adam Alter, *Irresistible: The Rise of Addictive Technology and the Business of Keeping Us Hooked* (New York: Penguin, 2017).

19 Eli Pariser, *The Filter Bubble: What the Internet Is Hiding from You* (New York: Penguin Books, 2012), 112–23.

20 Cathy O'Neil, *Weapons of Math Destruction: How Big Data Increases Inequality and Threatens Democracy* (New York: Crown, 2016), 150.

21 Fernando Nascimento and Anya Workman, "Rebalancing the Digital Convenience Equation through Narrative Imagination," *International Conference on Computer Ethics* 1, no. 1 (2023). https://soremo.library.iit.edu/index.php/CEPE2023/article/view/271.

22 Nascimento and Workman, "Rebalancing the Digital Convenience."

23 Daron Acemoglu and Simon Johnson, *Power and Progress: Our Thousand-Year Struggle Over Technology and Prosperity* (New York: PublicAffairs, 2023), 26.

24 Jonathan Donner, *After Access: Inclusion, Development, and a More Mobile Internet* (Cambridge, MA: MIT Press, 2015), 23.

25 Richard Ling has highlighted the implications for social cohesion of digital technologies' ubiquity: "the mobile phone has allowed us much more flexibility

when coordinating, or micro coordinating activities." Richard Ling, *New Tech, New Ties: How Mobile Communication Is Reshaping Social Cohesion* (Cambridge, MA: MIT Press, 2010), 185.

26 Ling, *New Tech, New Ties*, 185.

27 Elizabeth Marsh, Elvira Perez Vallejos, and Alexa Spence, "The Digital Workplace and Its Dark Side: An Integrative Review," *Computers in Human Behavior* 128 (March 1, 2022): 107–18.

28 Luciano Floridi summarized many aspects of digital ubiquity through the concept of "onlife" which captures the increasingly mixed nature of "online" and "offline" experiences: "it already makes little sense to ask whether one is online or offline when driving a car following the instructions of a navigation system that is updating its database in real-time. The same question will be incomprehensible to someone checking her email while traveling inside a driverless car guided by a GPS." Floridi, *The Fourth Revolution*. I will discuss the experience of digital technology in developing countries at a number of later points.

29 Lydia Saad, "Americans Have Close but Wary Bond with Their Smartphone," *Gallup*, June 20, 2022, https://news.gallup.com/poll/393785/americans-close-wary-bond-smartphone.aspx.

30 Saad, "Americans Have Close but Wary Bond with Their Smartphone."

31 Saad, "Americans Have Close but Wary Bond with Their Smartphone."

32 Crystal Hall, Eric Chown, and Fernando Nascimento, "A Critical, Analytical Framework for the Digital Machine," *Interdisciplinary Science Reviews: ISR* 46, no. 4 (October 2, 2021): 458–76.

33 Félix Cuadrado and Juan C. Dueñas, "Mobile Application Stores: Success Factors, Existing Approaches, and Future Developments," *IEEE Communications Magazine* 50, no. 11 (November 2012): 160–67.

34 Thomas Haigh and Paul E. Ceruzzi, *The Computer Is Everywhere and Nowhere* (Cambridge, MA: MIT Press, 2021), 396.

35 Krystal Hu, "ChatGPT Sets Record for Fastest-Growing User Base—Analyst Note," *Reuters*, February 2, 2023, https://www.reuters.com/technology/chatgpt-sets-record-fastest-growing-user-base-analyst-note-2023-02-01/.

36 Acemoglu and Johnson, *Power and Progress*, 13.

37 Mark Coeckelbergh, *AI Ethics* (Cambridge, MA: MIT Press, 2020), 86.

38 Bruno Latour, "On Technical Mediation," *Common Knowledge* 3, no. 2 (1994): 29–64.

39 Floridi, *The Fourth Revolution*, 25–32.

40 Floridi affirms that in many use cases, humans "are out of the loop entirely." Floridi, *The Fourth Revolution*, 35.

41 Ricoeur emphasizes metaphors' innovative and creative potential of "see one thing as another," which allows new meanings to emerge in language. Paul Ricoeur, *The Rule of Metaphor: Multi-Disciplinary Studies of the Creation of Meaning in Language* (Toronto: University of Toronto Press, 1978), 298.

42 Christian Stiegler, *The 360° Gaze: Immersions in Media, Society, and Culture* (Cambridge, MA: MIT Press, 2021), 7.

43 Eric Chown and Fernando Nascimento, *Meaningful Technologies: How Digital Metaphors Change the Way We Think and Live* (Ann Arbor: Lever Press, 2023).

44 Marianne Van Den Boomen, *Transcoding the Digital: How Metaphors Matter in New Media* (Amsterdam: Institute of Network Cultures, 2014).

45 Van Den Boomen, *Transcoding the Digital*, 12.

46 For example, Donald Schön extensively studied how metaphors can frame both design processes and public policy in different ways, leading to diverse ethical and social implications. Donald Schön, "Generative Metaphor: A Perspective on Problem-Setting in Social Policy," *Metaphor and Thought* 2 (1979): 137–63.

2

The Ethical Dimension of Digital Technology Development

Technical Production and Moral Action

In December 2020, technology professionals and the general public followed the troublesome news of Google's Artificial Intelligence Ethics group's Dr. Timnit Gebru's departure. The case of Dr. Gebru, who was co-leader of Google's Ethical AI team and a respected researcher, made headlines in major newspapers.[1] She left Google following her critique of its strategies for hiring minorities and the inherent biases present in modern artificial intelligence systems. The news had international repercussions,[2] which highlighted the importance of discussions surrounding the ethical consequences of new digital technologies, particularly new artificial intelligence algorithms. Dr. Gebru's decision to leave drew attention to recent nascent efforts by large technology companies to create internal departments and groups tasked with discussing ethical implications during the design and creation of digital artifacts. It is striking that in a field that has developed exponentially for the last seventy years, these ethical concerns have arisen only in the last decade. Such imbalance between the technical achievements of digital artifacts and architectures and the immaturity of discussions surrounding their ethical repercussions is the starting point for this chapter.

This chapter's objective is to investigate the relationship between technical production and moral action and to assess the extent to which ethical considerations are and potentially can be integrated into current practices of digital technologies' design, development, and deployment. In the first part of the chapter, I analyze the practice of software engineering in recent decades and the relationships between technique and ethics within this practice. This investigation will point to the growing awareness of the need for effective integration between ethical and technical discussions. The debate fostered by Dr. Gebru's case highlights the consequences of the rupture between moral action and technical production. Recent digital technologies, such as artificial intelligence-based systems, dramatically amplify these consequences given their ubiquity, increasing agency, and infrastructural role in current societies.

In the second part of this chapter, I consider Ricoeur's proposal regarding the relationship of technique and ethics more broadly through the concepts of *praxis* and *poiesis*, which, as usual in his work, puts together a creative perspective that places the two concepts, often considered contrary, in a productive tensional and dialectical relationship. This realization of the need to think about ethics and technique as two constantly coexisting dimensions of human action is one of this book's foundations. Ricoeur's analysis of the concepts of *praxis* and *poiesis* can provide a theoretical foundation for an integrative *Digital Poetics* approach that will be the guiding thread of this book. The urgency of creating a new Digital Poetics that recognizes and integrates ethical and technical dimensions through a form of configuration of digital artifacts will be—hopefully—one of the most significant contributions of this work. Let us begin, therefore, with a brief look at the practice of software engineering.

Software Engineering Practice

Hollywood loves portraying software development with over-the-top imagery. Think dimly lit rooms, hackers glued to glowing screens displaying lines of unreadable code racing down like waterfalls, and fingers flying across keyboards in a blur of lightning-fast typing. While these scenes might make for exciting visuals, they're far from the reality of most software development work. Developing the digital artifacts that shape so many dimensions of the world is much more than typing decryption algorithms frantically. While writing code in a programming language is often a core aspect of the software engineering practice,[3] several other steps are often overlooked in the public imagination but will be central to our analysis.

First, before starting the coding itself, it is necessary to define the purpose of the software. There is enormous variation in how this initial definition phase occurs in the development cycle,[4] and we will talk more about it in this chapter. Generally speaking, however, this phase is called "requirements specification." Next, developers need to plan the parts of the system and how they will communicate with each other; this is the software design phase. After the design, the coding itself takes place, which implements and translates the requirements and design into software components written in computer language. Finally, we can group a set of activities under the umbrella of software verification and validation, commonly referred to as software testing. This set of activities forms the software development cycle and has historically structured the practice of developing digital technologies.

Historical Aspects of Software Engineering Practice

In 2006, Barry Boehm, one of the most influential and prolific scholars of software engineering practice, shared his view of the history of software engineering in the twentieth and twenty-first centuries, using what he called a "Hegelian approach"[5] to show the cycles of thesis, antithesis, and synthesis. He proposed an arbitrary organization of the temporal units of his analysis into decades which, although debatable, allows some useful temporal references for our considerations about software engineering.

The first noticeable aspect of Boehm's retrospective history is the lack of references to ethics and morals. One can interpret this omission as an early indicator of a common disconnect between technical practices in engineering and their social and ethical implications. Boehm partly counterbalanced this omission by defining software engineering on the basis of the external implications of the practice. He describes it as "the application of science and mathematics to create software properties useful to people."[6] This definition is not only significant but also promising. Boehm clarifies that "useful to people" extends beyond computer science to include behavioral science, management science, and economics.[7] His explanation suggests a balanced approach in engineering: combining technical expertise (in mathematics and science) with a focus on practical utility.

Nevertheless, it is not immediately clear whether the knowledge of behavioral sciences, economics, and management science applies only to creating computational tools and means or to defining the ends that would characterize the use of such computational means. We must look at Boehm's analysis of software requirements to examine this question further. Although software requirements are the first artifacts generated in the software engineering lifecycle, they convey the end goals of the practice as the engineering objectives. Therefore, we need to better understand the relevance that the definition of requirements has in the definition of practice and what types of ends are captured by these requirements.

Boehm marks the beginning of software engineering practice in the 1950s with the development of three types of military-related systems: aircraft navigation, rocket launching, and radar control. The software development concept rose from the shadows of hardware development; hence, external specifications dictated by mechanical laws and operational specifications were the basis of software requirements. It is an interesting philosophical coincidence that in 1957 Ricoeur and Hannah Arendt wrote about the launch of Sputnik, the first artificial Earth satellite.[8] Both philosophers had broader intellectual goals in thinking about that technology, but their mentions of it represent precisely the kind of hardware and software that dominated the engineering of new technologies in the 1950s and 1960s.

Let me highlight two features of the software engineering practice's origin that persisted for several decades and are of paramount importance to our investigation. First, even with the gradual realization in the 1960s and 1970s of the more malleable nature of software development relative to hardware, the ideal of hardware engineering precision in meeting exact and predetermined standards and specifications remained a constant golden standard in software development practice. Boehm highlights, for example, the idea of formal methods whose objective was to create software that engineers could formally and mathematically validate and prove its correctness before making it available to external users. These objectives are still present in critical mission software such as aircraft autopilots or nuclear reactor controllers. However, in the early decades of software development, this correctness was the holy grail because most software was designed and used in these mission-critical use cases. In the 1960s, the delicious and not so mission-critical fruit-crushing games on your smartphone today were not on the list of possible software to be developed at scale.

Second, governments, public institutions, and universities were the primary software users in the first decades of development, which meant they had a clear understanding of their goals and requirements, and the developers' main concern was the adequate implementation of these external requirements. Historian Margaret O'Mara recently outlined in her book *The Code: Silicon Valley and the Remaking of America*,[9] the critical role of the military in the development of digital artifacts in the United States of the 1960s and 1970s.

Software development teams and users often had very similar backgrounds. Thus, the software requirements were specified in an intense collaboration between the representatives of companies that contracted the software development and its developers. It was also common that this relationship occurred within the same institutions, and an in-house software engineering team implemented the requirements of other departments of the same company. Both engineers and users inhabited similar and well-defined semantic universes. Users defined the goals of the software, and the engineers' primary task was to translate those objectives into machine code. Broadly speaking, users were responsible for the ethical aims, and developers were accountable for the technical efficiency of digital products.

With the growth of personal computers and the internet, users' profiles changed significantly in the 1980s and 1990s. In this period, two types of software development grew in parallel. On the one hand, users of tailor-designed software were primarily non-engineers, and the scope and impact of software engineering practice grew and expanded to small and medium-sized companies with new requirements for usability and interactivity. Boehm points out that there was now "the shift to user-interactive products with emergent rather than prespecifiable requirements."[10]

On the other hand, an explosion occurred in off-the-shelf software such as spreadsheets, text editors, database systems, and electronic games. These new software products represented a critical moment in the practice's evolution as they introduced a gap between the conception of the requirements (the software's main functionalities) and the software users. Thus, "software development tended to follow a bottom-up capabilities-to-requirements process rather than a top-down requirements-to-capabilities process."[11] This capabilities-to-requirements process meant that users needed to adjust to the capabilities offered by off-the-shelf products since the software was not created in the image and likeness of its users.

In the 2000s, the gap between developers and end users widened with the continued acceleration in the pace of information technology changes and the need to accelerate the time-to-market of software products. With new technologies of communication, productivity, and entertainment, software artifacts started to permeate and affect more and more the users' daily lives and personal activities. Additionally, industrial and commercial processes have come to rely heavily on digital systems for production control, logistics, human resources, and finance control.

Reflecting on this scenario of software development acceleration, Boehm concludes that models adapting the software to users' feedback (and complaints) *after* the system's deployment replaced the diligent models of gathering software requirements with the system's end users prior to implementation. Such new design models are risk-driven and adaptive. More and more software developers start to define the goals of the software, thus increasing the impact of their choices and decisions in the lives of an exponentially growing number of users. Boehm[12] expresses concern about the almost exclusive prioritization of productivity over other software development requirements, such as dependability, security, and privacy. Steve Jobs famously epitomized this process of developer autonomy in a 1998 interview in which he stated, "A lot of times, people don't know what they want until you show it to them."[13]

Aligned with our discussion of the peculiarities of digital technologies in the previous chapter, Boehm concludes his retrospective of the early decades of software engineering by predicting that the practice will meet new challenges given the growing demand for new "systems of systems" that must be developed in ever shorter periods. He maintains a language closely tied to the technical aspects of the practice with terms such as "capability verification and validation," "performance," and "dependability" and does not explicitly mention ethical dimensions and social implications as challenges for practice in the 2010s and beyond.

Before moving on to other sources dealing with the ethical aspects of software engineering, we highlight three aspects of Boehm's historical retrospective:

1. The practice was born and continued to be closely coupled to technical standards of excellence, whose heritage goes back to hardware engineering.
2. The primary actors defining software's requirements (objectives) changed substantially, from specialist customers to developers themselves, and with that, there was a considerable increase in developers' responsibility.
3. The social spheres impacted by software development grew exponentially, going from specialized and controlled environments such as the manufacturing facilities of space rockets to the creation of electronic mail applications used by millions daily.

Thus, over time, developers accumulated more responsibilities in defining the objectives of the software. This software has more comprehensive and sophisticated functions and impacts an ever more significant number of users and more complex aspects of society. On the other hand, the practice remains focused on primarily technical criteria of excellence with little emphasis on ethical issues and social impacts. Such a dichotomy has created a valley between the technical capacity of impacting multiple aspects of society and the required skills and methods to integrate ethical considerations of the objectives and consequences of these new technical capabilities.

Donald Gotterbarn recognized this gap between technique and ethics in his article on software engineering ethics for the 2002 second edition of the Encyclopedia of Software Engineering.[14] He argues that while there are diverse connotations for the term software engineering, the primary focus of the practice is the technical suitability of the product developed, and the core of the practice is optimization, scalability, reliability, and functional stability of digital artifacts. Carnegie Mellon University's Software Engineering Institute (SEI), one of the field's most respected and influential institutions, provided a significant example of this pragmatic view of the practice of software engineering. Its 1990 SEI Report on Undergraduate Software Engineering Education defined software engineering as a "form of engineering that applies the principles of computer science and mathematics to achieving cost-effective solutions to software problems."[15]

Gotterbarn expresses concern about the exclusive focus on technical aspects and highlights that in the 2000s, there was a growing awareness among software engineers and society regarding ethical and social aspects intrinsically related to technical solutions of software products. He argues that "a good software engineering decision requires an awareness of and careful consideration of both the technical and the ethical dimensions"[16] and supports that good software engineers need to have the technical competence to implement products accurately but also a set of moral values to guide their technical decisions. Throughout this book, such often neglected but inherent complementarity between technical capacity and ethical commitment interests us in particular as it defines one of the essential aspects of Digital Poetics.

Gotterbarn's text marks a significant advance in the awareness of the practice concerning the large-scale social repercussions of software product development but reveals two limitations of this emerging attention to ethical aspects. First, the main concern is negative, in the sense of preventing new applications from hurting users or society. The second, directly linked to the first, is that the proposed solution focuses on incorporating codes of ethics and professional conduct used almost as checklists for the underlying principle of "do no harm." Let's look at the implications of such limitations one at a time.

First, the exclusive focus on avoiding negative impacts of new software artifacts might create a mindset that minimizes exploring and prioritizing the potential positive impact of new technologies. Although necessary, this negative approach is not sufficient as it does not channel and structure developers' creative efforts toward the potential social benefits of the artifacts they are creating. Given the immense potential of new technological enablers, the focus on avoiding harmful consequences is essential but stifles a broad ethical reflection that prioritizes improving the common good. By neglecting this positive criterion for the design of digital technologies, the ethical analysis can become just an instrument to optimize economic attributes exclusively while minimizing the risks of adverse market consequences.

Second, the emphasis on the procedural use of ethical codes of conduct fails to address the new and unforeseen ethical dilemmas that technological advances generate outside of established principles. The emergence of brand-new ethical conundrums is increasingly evident and frequent in software development because of the exponential advancement in computational capacity and the ubiquity of digital artifacts that create usage scenarios difficult to capture by pre-established rules. The simple application of a professional code of conduct is insufficient, then, for two main reasons. The process of updating codes of conduct is typically much slower than the pace of digital innovations, and the generic nature of principles will always demand a complementary deliberative step that applies the principles to the uniqueness of the case.

The existing prioritization of a principles-based approach to ethical questions in software development began in the educational process of future software engineers. Even in undergraduate programs that tried to incorporate ethical discussions into their curricula, it tended to take place in isolated disciplines that concentrated on duty and rule-related aspects through a set of general principles. I am not suggesting that principle-based approaches are bad or unnecessary. On the contrary, such approaches are essential and, in many cases, extremely well structured, as in the case of the Code of Ethics and Professional Conduct of the Association for Computing Machinery.[17] However, although this Professional Code is one important part of the pedagogical process, it needs to be complemented by other forms of ethical reasoning that foster creative ways to consider ethical solutions for ever-new digital artifacts' design consequences. As Burton, Goldsmith, and Mattei point out, "there is a temptation to teach solely through the transmission of facts, rather than encouraging discussion and dissent. This approach, which many undergraduates have seen, can condition them to interpret what they learn in terms of an authority-based view of 'truth' that, in turn, leaves them unequipped to reason about situations involving no single correct answer or think cogently about ethical trade-offs."[18] Skiparn et al. also diagnose that "traditional ethics education for computer scientists may not include practical and timely training on how to weigh the consequences of their decisions."[19]

In addition to the gap between general principles and new and particular ethical dilemmas, there is also the problem of conflicting situations in applying two or more valid contradictory principles. For instance, several issues related to disinformation in social networks have raised ethical and legal dilemmas in evaluating conflicting principles such as freedom of expression and the protection of personal and institutional security.

After this brief overview of some key historical aspects of the software engineering practice, we turn our attention to the current state of affairs to consolidate a clear picture of the gaps between, on the one hand, the advancement and social impact of technological capabilities and, on the other hand, the training and practice of ethical consideration in the design of new digital artifacts.

The Current Landscape of Software Engineering Practice

Presently, ethical and social concerns about the impacts of new digital artifacts have gained strength in companies, academia, and the media. It is symptomatic that the most widely circulated newspapers have dedicated technology sections that almost daily bring stories about the impacts of technologies on data privacy issues, social polarization, misinformation, ransomware, cyber warfare, and cyberbullying, to name a few common topics. In early 2023, discussions about the ethical impacts of artificial intelligence reached all corners of society with the introduction of tools that use generative AI, such as those based on large language models. Over a million users installed the application ChatGPT[20] in less than a week, which, as previously noted, outshone the adoption speed of other digital artifacts such as Facebook and TikTok. In no time, the media and other public and private institutions recognized the potential impact of these new artificial intelligence methods on educational structures,[21] democratic systems, and the job market.[22]

There is a flourishing of research initiatives, study groups, conferences, and a plethora of publications exploring different possible approaches to integrating ethics in software development. I offer five sets of examples, drawn largely from academia but also from private companies, to which I will subsequently counterpose, through Ricoeur, my own proposal for Digital Poetics.

First, under the broader research and innovation umbrella, Wessel Reijers et al. published an extensive analysis of strategies for integrating ethical discussions into developing new artifacts. They highlight the "quickening pace of technological innovation and the ubiquitous use of the outcomes of R&I [Research and Innovation] processes in society."[23] The vast diversity of strategies led Reijers et al. to organize their findings in a tripartite categorization based primarily on the stage of development at which ethical interventions are focused. The first category, "ex-ante," encompasses initiatives that discuss ethical problems and propose solutions before product development begins. Initiatives "ex-ante" focus on emerging technologies and seek to explore ways technology enablers will impact the lives of people and society. Such enablers are not yet embedded in the development process of any product, so ethical discussions regarding their impact shape aspects of features and products that may eventually integrate these enablers. Consider, for instance, the design of surfaces that can render digital information, such as Liquid Crystal Displays (LCDs) and future holographic displays.[24] Regarding its potential impact, this is the perfect locus of ethical intervention, since it precludes the deployment of ill-conceived enablers that will be part of many digital artifacts in later stages of technology development. Assuming the intention to act ethically by developers and innovators, one of the issues of this approach is how to foresee unintended consequences of the enablers and implement the design in a way that minimizes possible harm. In chapter 3, we will discuss how the concepts of productive imaginative variations and fictional narratives efficiently operationalize this objective, which remains only an abstract ideal in many existing proposals.

Reijers et al.'s second category, "intra," contains methods that embed ethical deliberation into the technology design process.[25] The focus of "intra" interventions is to transform implementation steps to consider ethical attributes needed to inform technical decisions. One such scenario would be if the development team takes into account ethical criteria and possible biases when it designs a facial recognition algorithm. We will discuss how the concept of phronetic deliberation derived from Ricoeurian ethics contributes to this complex deliberation that involves both technical dimensions and moral principles applied to scenarios that are entirely new because of the very nature of creating new technological artifacts. Finally, Reijers et al. propose the "ex-post" category, which includes methods that assess the ethical implications of technological artifacts already in use.[26] An example of this intervention point is discussions on regulating content on social networks. At this stage, the delimitation of problems is more precise; however, some consequences prove irreversible, and changes may face practical, legal, and political obstacles. In addition to offering the fruitful categorization just presented, the bibliographic analysis by Reijers et al. carried out for the years between 1995 and 2015 confirms the growing concern with different approaches to integrating ethics and techniques in a period dominated by innovations related to computational functionalities.

A second relevant example for our discussion of current academic research on software engineering practices is the comprehensive study carried out by the Berkman Klein Center.[27] The Center mapped thirty-six documents with ethical principles for developing systems based on artificial intelligence, from governmental and nongovernmental groups of varying geographies. Published between 2017 and 2019, these documents proposing moral principles for the development of AI-based artifacts provide further evidence of the current concern with the relationship between ethics and technique in digital technologies. The Berkman Klein Center study also matters for our argument because the source documents analyzed and the final report produced are lists of principles such as accountability, explainability, and fairness that function as general criteria for moral evaluation. Yet, as we will see in our discussions on practical wisdom and narrative deliberation, they are in themselves insufficient for ethical decisions about ethical dilemmas that new digital technologies generate.

A third instance of the current academic discussion landscape regarding software engineering practices is associated with the pragmatic turn in the philosophy of technology. Going beyond the strict principle-based approach to issues in the development of emergent technologies in the 2000s, a group of philosophers of technology turned their attention to ethical issues coupled with the design of digital technologies. Instead of the more general and structural critiques offered by the Frankfurt School, such as Eric Fromm, by the later Heidegger, and by Ricoeur himself, the pragmatic turn of philosophers such as Langdon Winner, Don Ihde, Bruno Latour, Peter-Paul Verbeek, and others questioned the impact of each type of technology on people's lives and investigated technologies as integrated parts of human action and perception in the physical environment. This pragmatic approach prompted a new set of questions concerning the ethical dimensions linked to the design of new technologies. For instance, in 2008, Peter-Paul Verbeek wrote an article called "Morality in Design: Design Ethics and the Morality of Technological Artifacts,"[28] in which he elegantly defended the importance of ethical aspects in the construction of technologies that mediate human choices, perceptions, and actions and, therefore, are intrinsically linked to ethical issues. He highlighted that it was not enough to seek perfection in executing functional requirements, for one needed to investigate how technologies change the mediations between humans and the environment. Furthermore, in the same essay, Verbeek asserted the need to explore new technologies' unintended consequences by analyzing their future applications and how they will mediate human interactions after their deployment. In "Materialization Morality: Design Ethics and Technological Mediation,"[29] Verbeek also discussed the concept of "scripts," proposed by Madeleine Akrich and Bruno Latour, to analyze the ethical dimensions of technology design. Verbeek eloquently affirmed that engineers materialize morality. Technological artifacts are not neutral intermediaries but actively co-create the human being in the world. I will take up this concern and intuition later and propose that narrative imagination is a form of mediation analysis and a relevant source to address this genuine concern about the unexpected ways in which technologies affect human interactions, particularly users outside of groups garnering greater attention from developers of large technology companies.

In the same vein of integrating ethical issues into the early stages of developing new technologies, a fourth impactful theoretical proposal has arisen, also within the philosophy of technology, as a value-based theory. In "Value Sensitive Design: Shaping Technology with Moral Imagination,"[30] Batya Friedman and David G. Hendry reflected on the achievements of this approach in the two decades of its existence[31] and reaffirmed that "the importance of engaging human values in the technical design process cannot be stated strongly enough."[32] Friedman and Hendry described the theoretical and methodological aspects of the value-based theory and analyzed a rich set of techniques that can be applied to the design of new technologies in order to incorporate human values such as inclusivity, privacy, and justice, or as they summarize, what people hold important for their lives.

My fifth and final example comes from business and argues against an intentional relation between ethics and technology. In sharp contrast to the prioritization of preliminary deliberative processes based on human values applied during the conception phase of emerging technologies, some technology companies adopt a completely different approach and believe it is impossible to predict and inform the impacts of new technologies.[33] These companies decide to publish their artifacts to the general public and argue that society will self-regulate the use of digital artifacts based on individual and societal values as they interact with those artifacts. For example, generative AI models were made available to hundreds of millions of users quickly, without society or public representatives even minimally understanding the workings and implications of such technologies. While the companies that made these products capitalized their business with investments primarily focused on search engines and, consequently, advertisement, several social groups found themselves facing the arduous task of restructuring their practices due to the impacts of such digital artifacts. Large language models, for example, have created disruptions at various levels of the educational system, with teachers concerned about plagiarism and how to integrate these new technologies into coursework productively. As we discussed in chapter 1, digital materiality exponentially accelerates the impacts of the distribution of digital technologies. It creates enormous and worrying disruptions in fundamental structures for the functioning of institutions and social practices. The large technology companies' argument advocating for moderation by society after the fact is unsustainable and irresponsible. If anything, the societal disruptions brought about by the hasty deployment of large language model tools seem to make the need to productively integrate ethical discussions into the design of digital technologies even more compelling as a necessary aspect of technology development to minimize these disruptions to essential aspects of social and political fabrics.

This brief overview of the last decades of software engineering practice reveals some difficulties and challenges in integrating ethical concerns into the technical activities of the practice. Among them, I highlight four particularly impactful and relevant to our discussions in this chapter. First, a decoupling may occur between the requirements discussion of the purpose of the software and the design and implementation task of the software development lifecycle. This decoupling creates a false sense that technical decisions in software engineering practice are isolated from ethical and social implications. In its extreme manifestation, the requirements analysis (and therefore the discussion about objectives) is altogether alienated from the practice of software engineering, reducing the practice to the myth of a few genius coders who merely write computational algorithms.

The second difficulty is that even when the need to consider requirements is recognized, there might be a fixation on technical aspects of the technology and a depreciation or negligence of requirements related to the impact on people's lives by the artifacts. Thus, the exclusive priority of developers would be to make the software more efficient by performing its tasks optimally from the computational point of view, using the hardware resources well, and creating the most intuitive interface for users. All these requirements are valid and vital, but when they become the sole criteria for successful software design, requirements related to the ethical impacts of devices disappear from the developers' radar.

The third difficulty is that even when the need for discussing ethical requirements is acknowledged, there might be a focus on creating a catalog of moral principles that serve as checklists for verifying the developed functionality. While this is a positive and meaningful advance in integrating ethical attributes into technical processes, the principles are by design overarching and generic. This lack of specificity is a fundamental challenge in applying concepts to new technological enablers and artifacts. When dissociated from a deeper integration between technique and ethics, such lists of principles can become mechanisms to simply avoid future litigation or, in other cases, ways to establish merely an image of social responsibility for the companies that use them. Also, most guidelines are negative in the sense that they describe what developers should avoid doing. While this is a necessary condition, it is insufficient because it lacks a positive complement through which developers focus on goals linked to improving personal and social experiences.

The fourth difficulty arises in development groups that already integrate ethical analyses in defining the requirements of new technologies. In such groups, it is challenging to assess the unintended consequences of new functionalities and technology enablers, such as data sorting algorithms and communication protocols, even when carefully considering the guidelines captured in terms of general principles of good practice. The widespread use of digital technologies significantly broadens the range of contexts and areas they influence. As Verbeek has suggested, software developers need to devise ways to explore the future mediations of emerging technologies that will affect personal and social experiences after deployment.

To summarize, these four fundamental difficulties of software engineering practice concerning the integration of ethical aspects are as follows:

1. A disconnection between the practice of software development and the ethical and social consequences of technological artifacts;
2. An exclusive focus on technical requirements;
3. An overreliance on general deontological principles to solve specific ethical issues;
4. The difficulty of exploring unintended consequences when developing new features.

This synoptic view captures the problems I would next like to analyze. In the following section, I turn to the philosophy of Paul Ricoeur to outline elements of a possible conceptual framework for addressing each of these difficulties. I suggest that Ricoeur's analysis of action, ethics, and narrative offers a significant contribution to a deeper understanding of these challenges and opens up fruitful and complementary interactions with existing initiatives such as value-based theory and technological mediation analysis. I argue that Ricoeur's concepts can foster an organic integration of ethics and technology in software engineering practices and can become the basis of Digital Poetics, as I will pursue further in the following chapters.

Ricoeur's Take on the Relationship between Ethics and Technique

Praxis and *Poiesis*

One of the pathologies in software engineering practice that we identified in the previous section is the rupture between requirements and design. This break isolates the technical aspects of "how-to" algorithms, data structures, interfaces, and protocols from their application contexts and their intrinsic ethical implications. Often unnoticed by software engineers or intentionally concealed under the veil of alleged moral neutrality, this break opens the space for ethically debatable requirements to determine the design and development goals of new technologies. The isolation of technique from ethical aspects alienates designers and developers, enveloping them in a bubble of operational problems that neglect technological creations' ethical and social impacts. In this section, I explore how Ricoeur's philosophy of action challenges this separation between requirements and design at its root, showing that activities within practices always integrate, in varying degrees, "how to do it" and "why to do it," in other words, technical and ethical aspects.

We begin our analysis of how Ricoeur reframes the dialectic between ethics and technology with a brief detour through Aristotle's take on the subject. The Greek philosopher's thought greatly influenced Ricoeur's philosophy of action and established the semantic contours of the debate between technical production and moral action in the Western philosophical tradition. Aristotle defines and differentiates moral action (*praxis*) and technical production (*poiesis*) as follows:

> The class of things that admit variation includes both things and actions done. But making (*poiesis*) is different from doing (*praxis*) . . . Hence the rational quality concerned with doing is different from the rational part

> concerned with making. Nor is one of them part of the other, for doing is not a form of making; nor making a form of doing.[34]

Aristotle draws a clear distinction between *to poiein* (to make) and *to prattein* (to act). Within the semantic horizon of classical Athens, Aristotle proposes the distinction between an activity whose end is in itself (*praxis*) as opposed to other activities that are conditioned to necessity or external utility (*poiesis*), therefore, having an end other than itself. He also employs the term *techne* for the ability to make things well and *phronesis* for excellence in acting well according to the right reason (*kata orthon logon*). On the side of *poiesis* and *techne* are productive activities such as the construction of boats, the cultivation of fields, and even the writing of a tragedy. On the other side of *praxis* and *phronesis* are the ethical and proper political activities of deliberating well regarding the goals and means to achieve a good (or fulfilled) life (*eudaimonia*). As Gaëlle Fiasse has highlighted in her excellent analysis of Aristotle's *phronesis* and its relation to means and ends, Aristotle stresses his sharp differentiation between *praxis* and *poiesis* several times in the *Nicomachean Ethics* and other texts, as some ends are activities and others are works.[35] *Praxis* has its end in its activity, while artistic and productive activity (*poiesis*) has its end in a work extrinsic to the agent.

Further analyzing the context of this statement will open different perspectives for its integration into our inquiry. This conceptual description pertains to Book VI of the *Nicomachean Ethics*,[36] where Aristotle analyzes the specific functions of the mind concerning contingent things, that is, those things that are one way but could be another. He proposes two forms of intelligence that are primarily related to contingent affairs. One is aimed at building external things for the city, hence eminently productive, and another is aimed at thinking about what must be produced so that the person and the polis have a fulfilled life.[37] Aristotle writes a top-down analytical argument rendering precise the conceptual specificities of each concept. However, he is not, at this moment, discussing how these two distinct forms of intelligence can (and must) cooperate. And when one takes his statement separately or exclusively, we are left with the emphasis only on the conceptual distinction between *praxis* and *poiesis* and miss their practical interactions. I believe that interpretations following this reductive view of Aristotelian definitions are at the heart of some difficulties related to the sharp distinction between *praxis* and *poicsis* that we will discuss next.

Aristotle's description of the human endowed with practical wisdom (*phronesis*), the *phronimos*, provides a more comprehensive depiction of how *poiesis* and *praxis* can operate in tandem, even in deeply intertwined ways. Aristotle says that Pericles, his paradigmatic example of someone who has *phronesis* (and therefore excels in *praxis*), deliberates well with a view to the good (as a whole) of the polis, the political body.[38] But a deeper look at Pericles's actions, hailed by Aristotle, shows that his deliberation involves a decision that occurs in the linguistic form of orders and laws within the ancient Greek cultural and social horizon.

Pericles's participation in deliberations and decisions, guided by *phronesis*, at the *eklesia* (assembly) triggered a series of "makings" such as the construction of boats for the Athenian navy and the walls to protect the city. The strict conceptual separation between *praxis* and *poiesis* begins to be diluted in practice when one considers that the "for what" (*telos*), aimed at by *praxis*, and the "how" (*techne*), deployed by *poiesis*, intertwine in the details of the construction of the Athenian fleet that must take into account the degree of maneuverability that the vessel can have, how many crew members it can take, how much it costs for the polis, what other activities will not be carried out so that another ship can be built, and so on.

It seems naive to disregard the back-and-forth between the objectives of the good life for the polis and the details of carrying out tasks to achieve such goals, such as manufacturing ships for the Athenian fleet. For example, the ships' hydrodynamic and structural aspects could vary depending on the purposes of such vessels. It is equally reasonable to think that the deliberative calculations changed with the feedback of the execution of the tasks, creating a living interaction between *praxis* and *poiesis*.

To further understand the relationship between *praxis* and *poiesis*, we may break it into three complementary perspectives. The first is conceptual and aims at clarifying different ways of agency concerning contingent things. The differences between the two concepts are magnified at this level as the objective is to delineate each one's specificities. However, when we move from the conceptual to the practical sphere, a second perspective of differentiation emerges based on individual agents. Although Pericles is *phronimos* (excellent in *praxis*), he could also be an excellent boat builder (excellent in *poiesis*). It seems clear that *praxis* and *poiesis* can be complementary rather than mutually exclusive. But what interests us most is the third perspective of differentiation, that of the action. At this level, the excellencies of *praxis* and *poiesis* might be, and typically are, embodied by different agents, for instance, Pericles and an Athenian shipbuilder.

From the perspective of action, *praxis* and *poiesis* operate in tandem, so the polis's military or commercial goals inform the ship manufacturing process, and the manufacturing prompts reevaluations and adjustments on the goals and respective deliberations. Therefore, *praxis* and *poiesis* are practically intertwined and mutually implicated, even if they are still conceptually distinct. As we will see, it seems that the tradition of discussion and application of these concepts has privileged only the conceptual distinction and has forgotten the peculiar manner in which they are carried out in the intentional actions that affect the ways we act in the world.

Eric Schatzberg suggests in his history of the concept of technology that Aristotle's theoretical categorization had a fundamental influence on how *poiesis* has been detached from *praxis* and corroborated with an instrumental perspective on technology.[39] For different reasons that we do not need to address, the philosophical destinies of *phronesis* (and *praxis*) and *techne* (and *poiesis*) continued to diverge in the tradition of western thought.

John Wall agrees with Schazberg's assessment and lists Augustine, Aquinas, Kant, Nietzsche, and Habermas as other philosophical sources that, for various reasons, focused on the distinct aspects between poetic activities and moral decisions.[40] Wall's central thesis, of great interest to this section's objective, is that *phronesis (praxis)* involves a necessary element of poetics, making, or creativity. I suggest that beyond the poetic aspect of *praxis*, it is fundamental to recognize the dimension of moral deliberation in all *poiesis*. To this end, I present Ricoeur's argument of the mutual implication between *praxis* and *poiesis* in the context of practices and apply it to the case of software engineering.[41]

No Action Is Only *Poiesis* or Only *Praxis*

Ricoeur's take on *praxis* and *poiesis* provides us with a more productive application of these concepts to contemporary issues of digital technologies. In *Oneself as Another*, Ricoeur offers his systematic analysis of the relationship between *praxis* and *poiesis*. He begins by highlighting that despite the dichotomy we discussed earlier, Aristotle starts the *Nicomachean Ethics* by binding together *poiesis* (*techne*) and *praxis* in a parallel relationship with pursuing the ideal of the good: "Every art (*techne*) and every investigation, and likewise every practical pursuit (*praxis*) or undertaking, seems to aim at some good: hence it has been well said that the Good is That at which all things aim."[42] In Aristotle's famous opening of his ethics, *praxis* and *poiesis* cooperate in the essential process of aiming at the good proper to human beings although keeping their specific conceptual spheres.

At the beginning of his "Little Ethics" in *Oneself as Another* (chapters 7–9), Ricoeur takes this intuition and offers an ambitious, surprising, and welcome statement on the intrinsic relationship between *praxis* and *poiesis*: "If our analysis is correct, no action is only *poiesis* or only *praxis*."[43] As we explore this suggestion further, it will take us to the heart of Ricoeurian practical philosophy, in which narrative theory mediates between the theory of action and the theory of ethics, between description and prescription. Let us follow how Ricoeur organizes his argument through the concepts of coordination and subordination to support this new perspective on the relationship between *praxis* and *poiesis*.

The first notable conceptual movement is to consider *praxis* and *poiesis* from a comprehensive framework of meaningful actions that Ricoeur, using a concept extensively explored by Alasdair MacIntyre, calls "practices."[44] Thus, *praxis* and *poiesis* are thought within significant action structures such as farming, playing soccer, medicine, and, as we have been discussing, software engineering. Within practices, actions are combined in two dimensions, following two complementary heuristics. The first is one of coordination between actions in a linear way such that a given action X is done with a view to an action Y, which in turn aims at an action Z, and so on. This approach is close to the causal chains in analytical theories of action as in the philosophies of Anscombe[45] and von Wright.[46] The second heuristic is linked to a teleological subordination of actions that are no longer seen in a sequence relationship but in a nesting relationship. Thus, farming includes subordinate actions such as plowing, planting, and harvesting, and this decreasing hierarchy reaches basic actions such as pulling and pushing. Therefore, within a practice's universe of meaning, actions are interwoven not only in one dimension but simultaneously through coordination and subordination relations. According to Ricoeur, this same structure is applied to other levels of composite actions, such as a plan of life, but we will keep our discussion at the level of practices as it relates more directly to the problem and argument of this chapter.

Based on the complementarity of coordination and subordination in any meaningful actions within practices, we can then revisit Ricoeur's organic perspective on the intrinsic relationship between *poiesis* and *praxis*.

> From the viewpoint of linear coordination, the tie is more like Aristotelian *poiesis*, in which action has a result outside the agent in the sense that the result is external to the segment considered, to which the agent entrusts his or her power of acting; from the viewpoint of subordination, the tie is closer to Aristotelian *praxis*, in the sense that tilling the soil is done *pros to telos* in view of the end, while practicing the profession of farming is an action done "for itself," as long as the farmer does not question his choice of profession. If our analysis is correct, no action is only *poiesis* or only *praxis*.[47]

Ricoeur complements his analysis of this question, affirming that "actions must be *poiesis* in order to be *praxis*."[48] In constructing a ship, the craftsperson knows how to build the ship, and these practical skills aim at something exterior to oneself. However, if Ricoeur is correct, being a shipmaker is an end in itself, and every time the shipmaker makes (*poiesis*) a ship, he or she also acts as a shipmaker (*praxis*).

Ernst Wolff is another philosopher who, like Ricoeur, extensively discusses the relationship between technique and practice in the field of human actions. In his programmatic article, "'Technology' as the Critical Social Theory of Human Technicity,"[49] whose arguments he would later develop in his book *Between Daily Routine and Violent Protest: Interpreting the Technicity of Action*,[50] Wolff suggests an *integrated* hermeneutic and action-based method for understanding humans as technical actors, aiming to establish a foundation for a comprehensive theory of human agency and action's technicity.[51] The term "integrated" is employed in two ways: first, Wolff's research seeks to acknowledge and investigate the interconnectedness between technical and non-technical elements of human agency; second, Wolff's goal is to showcase how a theory of human technicity can effectively merge both descriptive and prescriptive aspects. Of particular interest to our discussion at hand is his reinterpretation of Aristotle's concepts of *praxis* and *poiesis*.

Wolff contends that the ethico-political (*praxis*) and the skillful (*poiesis*), as aspects of human abilities, possess several notable characteristics in common.[52] He maintains that the subtle differentiation between them does not represent a conflict; instead, it should be considered as harmonizing two mutually supplementary dimensions of action. To further his argument, Wolff comments that in Aristotle's definition of virtue, it is noteworthy that his action theory does not prohibit the creation of a comparable definition for skill.[53] In fact, the close relationship between skill and virtue may even be seen as inviting such a definition. The overlap between these concepts and the parallel definitions highlighting their similarities leads to Wolff's assertion that there exists only one disposition for decision-making *(hexis prohairetiké)* under the guidance of reason (*meta logou*). This disposition for rational choice-making operates in two similar yet distinguishable ways in the realms of *praxis* and *poiesis*, both of which should be perceived as deliberative striving *(orexis bouleutiké)*.[54]

Without needing to delve further into the complexities of Aristotle's portrayal of the relationship between *praxis* and *poiésis*, Wolff proposes as a result that *praxis* governs *poiesis*. In other words, virtuous action directs technical action through an intricate process of decision-making or choice, in the sense that virtue is involved with technical life as part of the essential intermediaries for its own realization. Given this argumentation, Wolff concludes that it is logical to view *praxis* and *poiesis* not as two distinct types of action but as two different aspects of all action, much like Ricoeur's conclusion that no action is purely either *praxis* or *poiesis*.[55]

In software engineering, the practitioner makes (*poiesis*) a sorting algorithm that coordinates with a user interface to display results on a digital screen.

However, this algorithm is part of a hierarchy of actions, such as creating a mobile application or a website. Through these higher-level actions, the sorting algorithm gains meaning and value within the software engineering practice that has a series of ends in itself (*praxis*). The developer is not just an algorithm maker, but he or she also acts as a software engineer. As another example, when a software engineering practitioner codes a screen that lists products, the developer is making (*poiesis*) something that has a purpose outside of the algorithm itself (e.g., it might be integrated into an e-commerce site), but this same action is the realization of the *praxis* of software engineering, which is an end in itself with intrinsic values.

Practices as meaningful frameworks of actions are key to supporting the intertwining of *praxis* and *poiesis*. Constitutive rules determine what it means to act within the practice. For instance, the game of chess has constitutive rules specifying that a given action is part of the practice, that is, that a given action counts as playing chess. For Ricoeur, these constitutive rules, which in themselves are ethically neutral, open up the evaluation of what shall be recognized as good within a given practice.[56] MacIntyre called standards of excellence the community-recognized values of the practice.[57] Thus, doctors agree on some values for the practice of medicine that become immanent goods of actions within the scope of the practice's sphere of meaning. This notion of immanent goods brings us back to the relationship between *praxis* and *poiesis*, as a particular surgical procedure is performed *(poiesis)* to unclog an artery, but it has an immanent good linked to the practice of medicine whose end in itself *(praxis)* is to save lives.

Software engineering as a practice also requires continuous and extensive discussion among its participants to determine its standards of excellence and immanent goods of the various actions within the practice. Given the intricate relationship between development phases, even if optimizing a machine learning algorithm seems like just *poiesis*, it must be placed in the context of subordination to other activities in the practice (such as the creation of financial applications) and evaluated against the standards of excellence that determine the practice of software engineering as a whole. Thus, the optimization of the algorithm is also *praxis*. The code developer never only makes a logical sequence of computational instructions but also acts according (or not) to the immanent good of the software engineering practice.

This conceptual framework raises several application questions, such as how and when standards for software engineering excellence are established? How are they integrated into the educational processes of future practitioners? How do these standards of excellence relate to broader societal goals, such as the flourishing of the common good? Such questions lie at the heart of our discussions in this book.

Exclusive Focus on Technical Requirements

In this section, we discuss the second difficulty raised in our brief descriptive analysis of practice regarding the exclusive focus on technical requirements, which is also related to the first question about how standards of excellence in software engineering are established. While the prior section showed the interrelation between *poiesis* and *praxis*, the next difficulty is the question of to what end these factors should be related. What is the excellence desired?

The difficulties of the software engineering practice that we are discussing in these sections are progressive in the sense that by overcoming one of them, the next ones become more apparent and more relevant. Thus, when we address the first difficulty of the dichotomy between *poiesis* and *praxis,* between designing new artifacts and considering this activity as part of a practice that has an end in itself and ethical implications, we are faced with the subsequent difficulty in deliberating about the practice's standards of excellence. So, assuming that all activities within the software engineering practice have unavoidable connections to standards of excellence set, explicitly or not, by its practitioners, one faces the next challenge to consider what are or should be the standards of excellence of the practice.

A fundamental risk is to reduce the definition of standards of excellence to a set of technical requirements that make the practitioner technically virtuous but allow him or her to ignore non-technical goals, such as how new technologies improve the quality of human life and the social and ecological environment. As we discussed in the first section, such reductionism is a danger ingrained in the highly technical roots of software engineering practice.

In *L'aventure technique et son horizon planétaire*,[58] a text from 1958, Ricoeur reflects on the growing relevance of technology in everyday life when the Cold War accelerated the technology and space race. Ricoeur suggests that technology has always had a place in human life; however, at that historical juncture, for the first time, it became an essential element of human self-awareness.[59] This new awareness made technology an essential element in defining what it means to be human and what is one's destiny. Ricoeur contrasts the accelerated technological development with two accompanying fragilities: the loss of meaning for practitioners and the problem of defining ends.[60]

Regarding the first fragility, Ricoeur correlates the productivity gains in industrialization with the loss of meaning in each person's work. The segmentation of production provided and motivated by new technologies makes it difficult to recognize personal meaning in professional tasks. Thus, if performing various tasks becomes less painful from a physical point of view, the result seems to be a growing lack of meaning in doing productive activities.

This disconnect between employees and the meaning of their activities, diagnosed by Ricoeur in the context of the increasing industrialization of the twentieth century, is equally problematic in software development today. To accelerate application development, designers create software components with isolated and specific functionality, which multiple applications reuse. For example, applications as diverse as a healthcare portal, online pizza sales, or payment for digital clothing for characters in video games might integrate the same digital payment component. Such componentization of digital artifacts generates two forms of dissociation between the development of new components and the final artifacts. On the one hand, component developers distance themselves from the end results and often ignore which products integrate their components. On the other hand, application developers integrate components they did not develop and may be unaware of each component's specific details or problems. Wessel and Helmer suggest that the modularization of digital technology development creates a gap between developers and the ethical repercussions of their activities. They defend that "the disaggregation of development responsibility also means the diffusion of responsibility for ethical outcomes."[61]

Concerning the second fragility Ricoeur raises, technological advancement also fosters the contradiction between the increased power to create and make things and an acute problem of choice and deliberation about the ends that guide such creations. Ricoeur defends that a meaningful end for technology must be derived from its service to humanity. In a 1975 text, he takes up the discussion of the purposes of technology in industrial society and points to the gap of meaning in the experiences of modern people. He invites us to question whether industrial society is enslaved to means without ends, which leads to another dimension of the loss of meaning that might be even more fundamental than the loss of other values, such as justice and liberty. He concludes by saying that the modern person "is perhaps more deprived of meaning than of justice."[62]

I believe Ricoeur's analysis of the multiple ways in which industrialized society and technological development have become dissociated from fundamental questions about meaning provides a conceptual framework for better understanding the second difficulty of software engineering linked to technical reductionism. One of the challenges for an organic, ethical proposal for developing new technologies is the recognition of this divorce between technical requirements and ethical implications in establishing standards of excellence in the practice of software engineering. Just as a doctor cannot be excellent in his or her practice without worrying about the good of public health, the development of new technologies must integrate into their standards of excellence criteria that transcend technical attributes and consider the common good. This demand leads us to the next difficulty of software engineering practice that we discuss in the following section.

Overreliance on Deontological Principles to Solve Ethical Issues

As we noticed in the brief historical review of software engineering practice, the last few decades have seen a series of positive and relevant efforts toward establishing ethical principles for developers of new technologies. The excellent Code of Ethics and Professional Conduct of the Association for Computing Machinery[63] and the various documents of ethical principles related to artificial intelligence gathered by the Principled AI report of the Berkman Klein Center[64] are examples of initiatives that define technical and non-technical criteria, proposing standards of excellence for the practice of software engineering.

Such initiatives overcome the first two difficulties of the software engineering practice, as they recognize the intrinsic relationship between making digital technologies and acting morally. Furthermore, they also transcend merely technical criteria, including social and ethical dimensions in their principles. However, although code of conduct documents or a list of moral principles to be respected by developers of new digital technologies are necessary for establishing meaningful standards of excellence, they are not sufficient for at least three reasons.

First, the pace of new technology development is much faster than the elaboration of codes of conduct, professional guidelines, and moral principles, which require public deliberation and, in some cases, legal bureaucratic procedures to become legislative instruments. Second, the necessary generality of principles always leaves out the ethical issues related to the specific contexts of design and use of emerging technologies. The particular contexts of technology use are critical to decisions about implementation, deployment, use, and regulation, and they are never exhausted in the general descriptions of principles-based documents. Finally, there are cases where certain digital artifacts satisfy one principle but not others, creating a dilemma in the application of the proposed principles. For example, as mentioned earlier, in the case of infectious disease tracking applications, there is an inherent tension between principles that prioritize public health and others that prioritize user privacy.[65]

Thus, the development of digital technologies revives the well-known dilemma of applied ethics between deontological principles and teleological contextual deliberations. And again, Ricoeur's practical philosophy offers valuable insights to overcome this paradox through the elegant proposal of his "Little Ethics" that organically integrates the need for general moral principles with deliberative forms that address the difficulties inherent in the contextual application of general principles, particularly in a practical region as dynamic as digital technologies.[66]

As noted in *Critique and Conviction*, Ricoeur relates the idea of hermeneutic application to *phronesis* when discussing the importance of practical wisdom in his "Little Ethics": "What is it to solve an entirely novel ethical-practical problem?—this is the problem of practical wisdom, which I connect to the hermeneutics of 'application' under the aegis of Aristotelian *phronesis*."[67] Ricoeur's attention is focused on hard cases in the spheres of law, medicine, or everyday life, but his remark is a fabulous indication of the affinity between phronesis and the ethical issues associated with new technologies. Technological innovations create unprecedented scenarios that require phronetic deliberation, which applies principles and rules to particular situations through ethical imagination.

Ricoeur recontextualizes and reshapes the concept of practical wisdom by integrating it into a deliberative arc that includes the aim at good objectives and the deontological test proposed by Kant's universalization test. The premise is that one starts the ethical thought process willing to act well; there is a goal to do what is good. However, there is also always the need to test good intentions with the universalization rule that provides a logical heuristic to ensure we are respecting others while aiming at good objectives. But the logical rules face the complex problems of application when one needs to make particular decisions guided and filtered by general principles. There are ethical aporias created by an ascending and descending cycle of creating and applying moral rules. Different maxims may pass the deontological test and become acceptable as moral rules in the ascending direction from maxim to rules. But when a complex new ethical dilemma arises, the application of the rule in a different direction—a descending one—from the rule to the concrete situation becomes problematic. It is, then, when the specificity of the case and the respect for the people involved are at issue that it may unveil the conflict among different logically valid moral rules. That's the new context in which Ricoeur locates practical wisdom, which is not an isolated mode of ethical thinking but a necessary step toward the just application of moral principles in the process of doing what is good. Practical wisdom addresses the conflict that arises not from the randomness of an amoral decision but from the practical quandary derived from the application of conflicting principles that passed the logical test of respect for others in the ascending direction from maxims to principles. As is typical of Ricoeur's works, the strength of the reconceptualization of practical wisdom stems from a productive dialectic between universal principles and contextual application.

As we will explore further in the following chapters, the critical practical wisdom Ricoeur proposes is eminently creative. Faced with the aporia created by applying conflicting deontological concepts or general principles not fully adjusted general principles to the particular situation, practical wisdom is an activity of creative imagination that composes an ethical deliberation that betrays the general principles to the smallest extent possible and prioritizes respect for each person. Thus, the discussion that we will have in the following chapters about ethical deliberation based on creative imagination is an attempt to respond practically to the aporias and inadequacies of a model of ethical deliberation based on general rules within software engineering practice. In software engineering, a developer may decide to apply his or her skills to create technologies that foster the common good. The professional principles and guidelines provide a deontological framework to test and evaluate choices according to general ethical constraints. However, applying these general principles to specific design and implementation decisions requires a kind of moral judgment akin to practical wisdom.

The next section discusses the fourth difficulty of integrating ethical principles into software engineering practice related to challenges of unintended consequences given the multiplicity of uses of digital technologies.

Accounting for Unintended Consequences

In 1996, the United States Congress enacted Section 230 of the Telecommunications Decency Act.[68] The explicit premises for this legislation included the enormous potential for education and information sharing that the emerging expansion of internet access represented. The findings also highlighted the potential benefit of Internet communications for strengthening democratic debate, cultural diversity, and other intellectual activities. Based on these premises, lawmakers effectively allowed digital communication platforms operating on the internet to be considered content intermediaries rather than content creators. As mere intermediaries, communication platforms such as social networks did not become liable for the content that passed through their computers. Additionally, the legislation also provided for a "Good Samaritan" clause, which allowed service providers, acting in good faith, to remove objectionable content.

Our advantaged view of the past reveals how the reality of the content and use of digital communication platforms has diverged from the premises set forth in Section 230 of the Telecommunications Decency Act of 1996. The exponential growth in the use of digital platforms has demonstrated their potential to negatively impact political discourse through polarization amplified by algorithms that select the information that reaches users. The content of the most popular social networks seems far removed from the privileged forums for cultural exchange and the promotion of educational information predicted by legislators in the 1990s. However, to be fair in our assessment of Section 230, in analyzing the actual technological conditions in 1996, it would have been impossible to predict all the technological, social, and economic changes that have transformed the dial-up connections of the 1990s into fiber-based broadband connections capable of supporting the load of remote connections imposed by the pandemic of 2020.

The problems of unintended consequences of digital technology do not just happen on massive scales and over decades, as in the case of the internet. For example, several devices that monitor and act remotely through digital networks aim to bring their users comfort and security. However, Wi-Fi cameras, smartphones, smart doorbells, and smart thermostats are often taken advantage of by abusers[69] to control and intimidate their victims, who become hostages of behaviors mediated and amplified by digital technologies, whose functionalities are often opaque to the victims. Similarly, social networks allow online connections between geographically dispersed people. Still, they can unintentionally diminish the ability to establish and cultivate meaningful face-to-face relationships.[70]

Examples of unintended side effects of digital technologies are diverse, affect different forms of agency, and are particularly troubling when they affect more vulnerable audiences such as children and youth. The underlying issue for us is that even with the establishment of ethical principles, the speed of technological development and deployment requires a special exercise in exploring unintended consequences. The fundamental question is how to expand ethical deliberation on new technological decisions to consider more social contexts affected by technologies, especially those of more vulnerable social groups.

As we have mentioned, Peter-Paul Verbeek discussed the concept of "technological mediation" and how technologies shape human actions and influence the relationship between human beings and the world.[71] He argues that developers need to study how artifacts will mediate the experiences and actions of their users. One of the great difficulties of this investigation is that, using Don Ihde's terminology, technologies are multistable,[72] meaning their mediations are not the same for all usage contexts; the impact of technologies changes according to specific interactions between the technology and its users. To meet the challenge of the multistability of technological innovations, Verbeek proposed two strategies for analyzing mediation. In the first strategy, designers should imagine the mediating role that artifacts will have in different contexts of use and adjust design aspects based on the anticipations from this imaginative process. Verbeek proposed a second mediation analysis strategy based on the constructive technology assessment (CTA) method, which aims to involve all relevant stakeholders in the technology design process. CTA prescribes a structure for stakeholders to meet and discuss the functionalities of the future technology and build consensus on how the technology should be designed. Mediation analysis extends the initial analysis of functionality to consider not only the attributes of the artifacts but more importantly the mediations involved in order to promote desired mediations and avoid undesired ones.

Verbeek's intuition about the materialization of ethical principles through digital technologies and the need for mediation analysis of future technologies fits perfectly with the central arguments of this chapter. However, a theoretical framework that situates digital technologies within the narrative and ethical structure of human action, as proposed by Ricoeur, considerably enhances the mediation analysis in at least two important dimensions that I will discuss next. First, Ricoeur's narrative theory contributes significantly to thinking about practical directions that can minimize the negative impacts of unpredictability for digital technological innovations. Ricoeur suggests that narratives can function as ethical laboratories[73] in which we experience the effects and consequences of certain actions and choices through narrative representations. Narratives allow for the consideration of various voices, perspectives, and contexts in which technologies have implications, creating a personal distance that decentralizes the egoistic perspective, allowing alterity to permeate the ways we relate to technologies. For example, considering narratives, developers may explore different contexts in which certain types of digital artifacts will be used.

Thus, the investigation of mediations impacted by new digital technologies can be carried out through a narrative imagination exercise that places technologies in narrative contexts with different characters, settings, and plots. Narratives can and should represent different cultural, social, and economic perspectives from which digital technologies take on different shapes and mediate experiences and actions differently. In the next chapter, I present a comprehensive framework based on Ricoeur's narrative theory to address our last concern regarding the integration of ethics and poetics in software engineering practice.

A second enhancement of Verbeek's mediation analysis regards the involvement of multiple stakeholders in the design process. In the next chapter, we develop Ricoeur's intuition about the plurality of ethical deliberations. When discussing the application of moral principles to particular cases in *Oneself as Another*, Ricoeur mentions the possibility and benefits of an ethical decision emerging from a group of people through public debate. Using the figure of the *phronimos* (one who possesses practical wisdom for good ethical deliberations), Ricoeur asserts: "The *phronimos* is not necessarily one individual alone."[74] This suggestion by Ricoeur is a key intuition in our pursuit of broadening the technology development deliberative process through a narrative framework capable of minimizing unintended consequences through a diverse and inclusive group of deliberators.

This chapter presented a brief retrospective of software engineering practice, which is at the heart of developing new digital technologies. In this trajectory, we highlighted some cardinal difficulties for the consistent and productive integration of ethical and technical aspects. We have ordered these difficulties in a very intentional way so that by addressing one, the next becomes more apparent and relevant. Thus, initially, (1) the biggest challenge is to recognize that there is an ethical dimension in every software engineering activity. Next, (2) one must identify social and ethical requirements, not just technical ones, as essential to excellence in practice. Having overcome these two difficulties, practitioners face (3) the search for ways to deliberate ethically in their technical activities. We have seen how several current initiatives focus ethical deliberation on discussing general moral principles that must be supplemented by a kind of practical rationality for the application and constant adjustment, particularly in a highly dynamic practice such as the development of digital technologies. The practitioner who recognizes the need for this contextual deliberation based on general principles encounters the fourth difficulty: how to (4) explore the different application contexts of new technology to minimize unintended consequences, maximize inclusivity, and promote the common good.

For each of these difficulties, we briefly pointed to contributions from Ricoeur's philosophy that will form the basis for the development, in the next chapter, of a conceptual framework for ethical deliberation on digital technologies. In particular, instructed by Ricoeur's theory of action, we emphasized the importance of considering developers as members of a practice. Based on this concept of practice, we extensively explored the intrinsic and necessary relationship between the ethical and technical aspects of developing technologies.

Two guiding questions emerge from this reflection on the ethical dimension of developing new digital technologies. The first is what deliberative process can address the need to integrate ethics and technology, given the speed of development of digital technologies and the ubiquity of their impact, which generates multiple scenarios of use and countless contexts of application? This issue, and to some extent this entire chapter, originates from the optimistic premise that the developers of new technologies freely choose ethical objectives in line with the common good. Starting from this premise, the fundamental difficulties are linked to the transformation of an ideal of a good life, mediated by moral principles, into technologies that realize this ideal. This optimistic approach is necessary but needs to be complemented so that a more robust ethical proposal takes into account cases in which the values of the developers or technology development companies are not aligned with the ideas of the fulfilled life of the society and communities impacted by the new technologies. Such possible misalignment of ethical goals between technology companies and the common good poses the second guiding question for the following chapter: Who shall be the actors in this deliberative process?

In this chapter, we have focused on the developers of new technologies. Still, as we have seen, even assuming an optimistic perspective, the digital malleability and customization that exponentialize the possibilities of unintended consequences suggest that other actors also need to participate in the deliberative process of Digital Poetics. Furthermore, the increasing impact of digital technologies on core societal and individual values requires a framework that accounts for multiple deliberative actors—not only developers—who can exercise their agency in the Digital Poetics and foster productive forms of integrating ethical values with emerging technological capabilities.

In the next chapter, we will present narrative deliberation as a suitable process to cope with the difficulties raised in this chapter and suggest expanding this deliberative process to the spheres of users and institutions that complement the deliberative efforts of developers in promoting digital technologies that prioritize ethical goals.

Notes

1 Nitasha Tiku, "Google Hired Timnit Gebru to be an Outspoken Critic of Unethical AI. Then She Was Fired for it," *The Washington Post*, December 23, 2020, https://www.washingtonpost.com/technology/2020/12/23/google-timnit-gebru-ai-ethics/. Cade Metz and Daisuke Wakabayashi, "Google Researcher Says She Was Fired Over Paper Highlighting Bias in A.I.," *The New York Times*, December 3, 2020, https://www.nytimes.com/2020/12/03/technology/google-researcher-timnit-gebru.html.

2 Le Monde, "Le licenciement polémique de Timnit Gebru, qui travaillait chez Google sur les questions d'éthique liées à l'IA," December 7, 2020, https://www.lemonde.fr/pixels/article/2020/12/05/apres-le-licenciement-d-une-chercheuse-noire-et-militante-google-somme-de-s-expliquer_6062328_4408996.html. Julia Carrie Wong, "More than 1,200 Google Workers Condemn Firing of AI Scientist Timnit Gebru," *The Guardian*, December 7 2020, https://www.theguardian.com/technology/2020/dec/04/timnit-gebru-google-ai-fired-diversity-ethics.
3 With the advent of large language models, even certain aspects of coding in programming languages have been automated.
4 For a detailed overview of these phases, see Roger Pressman's software engineering manual, which has become very popular in computer science courses. Roger S. Pressman and Bruce R. Maxim, *Software Engineering: A Practitioner's Approach* (New York: McGraw-Hill Education, 2019).
5 Barry Boehm, "A View of 20th and 21st Century Software Engineering," In *Proceedings of the 28th International Conference on Software Engineering*, 12–29 (New York: Association for Computing Machinery, 2006), 12–13.
6 Boehm, "A View of 20th and 21st Century Software Engineering," 12.
7 Boehm, "A View of 20th and 21st Century Software Engineering," 12.
8 Paul Ricoeur, "L'aventure Technique et Son Horizon Planétaire," *Autres Temps* 76, no. 1 (2003): 67–78. Hannah Arendt, *The Human Condition*, 2nd Edition (Chicago: University of Chicago Press, 2013), Kindle Edition. Foreword.
9 Margaret O'Mara, *The Code: Silicon Valley and the Remaking of America* (New York: Penguin, 2020), 23.
10 Boehm, "A View of 20th and 21st Century Software Engineering," 18.
11 Boehm, "A View of 20th and 21st Century Software Engineering," 18.
12 Boehm, "A View of 20th and 21st Century Software Engineering," 20. See also 1999 PITAC Report, "The IT industry spends the bulk of its resources, both financial and human, on rapidly bringing products to market." [123]. Recognition of the problem is increasing. ACM President David Patterson has called for the formation of a top-priority Security/Privacy, Usability, and Reliability (SPUR) initiative [119].
13 "Steve Jobs," in *Oxford Essential Quotations*, edited by Susan Ratcliffe (Oxford: Oxford University Press, 2016), https://www.oxfordreference.com/view/10.1093/acref/9780191826719.001.0001/q-oro-ed4-00005922.
14 Donald Gotterbarn, "Software Engineering Ethics," in *Encyclopedia of Software Engineering*, edited by J. J. Marciniak (New York: Wiley, 2002).

15 Gary Ford, "990 SEI Report on Undergraduate Software Engineering Education," in *Carnegie-Mellon University* (Software Engineering Institute, 1990), https://citeseerx.ist.psu.edu/document?repid=rep1&type=pdf&doi=9b7e519bf1f9dc772a4320414bb3e1344774fda2.

16 Gotterbarn, "Software Engineering Ethics," 1–2.

17 D. W. Gotterbarn, B. Brinkman, C. Flick, and M. S. Kirkpatrick, "ACM Code of Ethics and Professional Conduct Preamble," Accessed December 21, 2023, https://dora.dmu.ac.uk/bitstream/handle/2086/16422/acm-code-of-ethics-and-professional-conduct.pdf?sequence=1.

18 Emanuelle Burton, Judy Goldsmith, and Nicholas Mattei, "How to Teach Computer Ethics Through Science Fiction," *Communications of the ACM* 61, no. 8 (July 2018): 59.

19 Michael Skirpan, Nathan Beard, Srinjita Bhaduri, Casey Fiesler, and Tom Yeh, "Ethics Education in Context: A Case Study of Novel Ethics Activities for the CS Classroom," in *Proceedings of the 49th ACM Technical Symposium on Computer Science Education* (New York: Association for Computing Machinery, 2018), 940–45.

20 Hu Krystal, "ChatGPT Sets Record for Fastest-Growing User Base—Analyst Note," *Reuters*, February 2, 2023, https://www.reuters.com/technology/chatgpt-sets-record-fastest-growing-user-base-analyst-note-2023-02-01/.

21 Will Douglas Heaven, "ChatGPT Is Going to Change Education, Not Destroy It," *MIT Technology Review*, April 6, 2023, https://www.technologyreview.com/2023/04/06/1071059/chatgpt-change-not-destroy-education-openai/.

22 Lydia DePillis and Steve Lohr, "Tinkering With ChatGPT, Workers Wonder: Will This Take My Job?" *The New York Times*, March 28, 2023, https://www.nytimes.com/2023/03/28/business/economy/jobs-ai-artificial-intelligence-chatgpt.html.

23 Wessel Reijers, David Wright, Philip Brey, Karsten Weber, Rowena Rodrigues, Declan O'Sullivan, and Bert Gordijn, "Methods for Practising Ethics in Research and Innovation: A Literature Review, Critical Analysis and Recommendations," *Science and Engineering Ethics* 24, no. 5 (October 2018): 1439.

24 Chenliang Chang, Kiseung Bang, Gordon Wetzstein, Byoungho Lee, and Liang Gao, "Toward the Next-Generation VR/AR Optics: A Review of Holographic Near-Eye Displays from a Human-Centric Perspective," *Optica* 7, no. 11 (November 20, 2020): 1563–78.

25 Reijers, et al., "Methods for Practising Ethics in Research and Innovation," 1451–52.

26 Reijers, et al., "Methods for Practising Ethics in Research and Innovation," 1452–54.

27 Jessica Fjeld, Nele Achten, Hannah Hilligoss, Adam Nagy, and Madhulika Srikumar, "Principled Artificial Intelligence: Mapping Consensus in Ethical and Rights-Based Approaches to Principles for AI," *Berkman Klein Center Research Publication*, no. 2020-1 (January 2020), https://doi.org/10.2139/ssrn.3518482.

28 Peter-Paul Verbeek, "Morality in Design: Design Ethics and the Morality of Technological Artifacts," in *Philosophy and Design: From Engineering to Architecture*, edited by Peter Kroes, Pieter E. Vermaas, Andrew Light, and Steven A. Moore (Dordrecht: Springer Netherlands, 2008), 91–103.

29 Peter-Paul Verbeek, "Materializing Morality: Design Ethics and Technological Mediation," *Science, Technology & Human Values* 31, no. 3 (May 1, 2006): 361–80.

30 Batya Friedman and David G. Hendry, *Value Sensitive Design: Shaping Technology with Moral Imagination* (Cambridge, MA: MIT Press, 2019).

31 See Batya's seminal article published in 1996: Batya Friedman, "Value-Sensitive Design," *Interactions* 3, no. 6 (December 1, 1996): 16–23. See also, for example, Batya Friedman and Helen Nissenbaum, "Bias in Computer Systems," *ACM Transactions on Information and System Security* 14, no. 3 (July 1, 1996): 330–47. Helen Nissenbaum, "Protecting Privacy in an Information Age: The Problem of Privacy in Public," *Law and Philosophy* 17, no. 5/6 (November 1998): 559.

32 Batya and Hendry, *Value Sensitive Design*, 167.

33 Emad Mostaque, "Stable Diffusion Launch Announcement—," *Stability AI*, August 10, 2022. https://stability.ai/blog/stable-diffusion-announcement.

34 Aristotle, *Nicomachean Ethics* (Indianapolis: Hackett Publishing, 2019), VI, iv, 1.

35 Gaëlle Fiasse, "Aristotle's φρόνησις: A True Grasp of Ends As Well As Means?" *The Review of Metaphysics* 55, no. 2 (2001): 323–37.

36 For technical discussions on Aristotle's descriptions of intellectual virtues, see, for instance, C. D. C. Reeve, "Aristotle on the Virtues of Thought," in *The Blackwell Guide to Aristotle's Nicomachean Ethics*, edited by Richard Kraut (New York: John Wiley & Sons, 2008), 199–218; and Michael Pakaluk, ed., "Thinking-Related Virtue (Nicomachean Ethics, Book 6)," in *Aristotle's Nicomachean Ethics: An Introduction*. Cambridge Introductions to Key Philosophical Texts (Cambridge: Cambridge University Press, 2005), 206–32.

37 Aristotle, *Nicomachean Ethics*, VI, v, 1. "Now it is held to be the mark of the prudent man to be able to deliberate well about what is good and advantageous

for himself, not in some one department, for instance what is good for his health or strength, but what is advantageous as a means to the good life in general."

38 Aristotle, *Nicomachean Ethics*, VI, v, 5—"Hence men like Pericles are deemed prudent, because they possess a faculty of discerning what things are good for themselves and for mankind; and that is our conception of an expert in Domestic Economy or Political Science."

39 Eric Schatzberg, *Technology: Critical History of a Concept* (Chicago: University of Chicago Press, 2018).

40 John Wall, "Phronesis, Poetics, and Moral Creativity," *Ethical Theory and Moral Practice: An International Forum* 6, no. 3 (September 1, 2003): 317–41.

41 Besides Ricoeur's arguments based on the meaningfulness of actions within practices, other commentators have noticed the logical problems of a strict separation of *praxis* and *poiesis*. For instance, Gyorgy Markus examines several aporias describing all actions as either *praxis* or *poiesis*. He concludes:

> Against the Aristotelian dictum, according to which "neither is acting (praxis) making, nor is making acting," the fact is that many of our makings are at the same time doings, actings in the narrow sense, most of our actings are simultaneously makings of one sort or another. The attempt to keep these two classes strictly separate as distinct types of purposeful human activity leads not only to logical incoherence (the constant shift in the employment of the terms from the objective to the subjective plane and vice versa; the unclarities regarding the teleological structure of action etc.), but also stimulates a search for pure praxis, for praxis which is not and cannot ever be poiesis and which then turns out not to be action any more. (Gyorgy Markus, "Beyond The Dichotomy: Praxis and Poiesis," *Culture, Science, Society* 43 (2011): 37–57. Brill)

42 Aristotle, *Nicomachean Ethics*, I, I, 1094al-3.

43 Paul Ricoeur, *Oneself as Another* (Chicago: University of Chicago Press, 1994), 176.

44 Alasdair MacIntyre, *After Virtue* (Notre Dame: University of Notre Dame Press, 1981), 12. Ricoeur, *Oneself as Another*, 176.

45 G. E. M Anscombe, *Intention* (Cambridge, MA: Harvard University Press, 2000).

46 Georg Henrik von Wright, *Explanation and Understanding* (Ithaca: Cornell University Press, 2004).

47 Ricoeur, *Oneself as Another*, 176.

48 Paul Ricoeur, *Soi-même comme un autre* (Paris: Seuil, 1990), 207. "Elle doit être poièsis en vue d'être praxis." This sentence in the original was missed in the English translation.

49 Ernst Wolff, "'Technology' as the Critical Social Theory of Human Technicity," *Journal of Philosophical Research* 41 (October 6, 2016): 333–69.

50 Ernst Wolff, *Between Daily Routine and Violent Protest: Interpreting the Technicity of Action* (Berlin-Boston: De Gruyter. 2021), https://library.oapen.org/handle/20.500.12657/50227.

51 Wolff, "Technology as the Critical Social Theory of Human Technicity," 334.

52 Wolff, "Technology as the Critical Social Theory of Human Technicity," 338–42.

53 Wolff, "Technology as the Critical Social Theory of Human Technicity," 341.

54 Aristotle, *Nicomachean Ethics*, 1140a3-5

55 Wolff, "Technology as the Critical Social Theory of Human Technicity," 342.

56 Ricoeur, *Oneself as Another*, 155.

57 MacIntyre, *After Virtue*, 218.

58 Paul Ricoeur, "L'aventure technique et son horizon planétaire," *Esprit*, 2006, https://doi.org/10.3917/espri.0603.0098.

59 Ricoeur, "L'aventure Technique et Son Horizon Planétaire," 2006, 95–96.

60 Ricoeur, "L'aventure Technique et Son Horizon Planétaire," 2006. 105.

61 Max Wessel and Nicole Helmer, "A Crisis of Ethics in Technology Innovation," *The Next Age of Disruption*, 2021, https://doi.org/10.7551/mitpress/13768.003.0008.

62 Paul Ricoeur, *Political and Social Essays* (Athens: Ohio University Press, 1975), 172.

63 Gotterbarn et al., "ACM Code of Ethics and Professional Conduct Preamble."

64 Fjeld et al., "Principled Artificial Intelligence: Mapping Consensus in Ethical and Rights-Based Approaches to Principles for AI."

65 Federica Lucivero, Nina Hallowell, Stephanie Johnson, Barbara Prainsack, Gabrielle Samuel, and Tamar Sharon, "COVID-19 and Contact Tracing Apps: Ethical Challenges for a Social Experiment on a Global Scale," *Journal of Bioethical Inquiry* 17, no. 4 (December 2020): 835–39.

66 Ricoeur's "Little Ethics" structures study 7, 8, and 9 of *Oneself as Another*, culminating the long arc of the previous chapters that explore human beings' capacities to speak, act, narrate, and recognize themselves as ethical subjects. In study 7, Ricoeur emphasizes the ethical desire to pursue a fulfilled life with and for others in just institutions. In study 8, Ricoeur introduces a deontological

component into the trajectory of the "Little Ethics" that tests ethical ideals based on the formal procedures proposed by Kant to guarantee respect for other people. Finally, in chapter 9, Ricoeur recognizes that the formal procedures of the deontological approach generate aporias in applying moral rules, which again demand recourse to the ideal of the good life and to practical wisdom, which makes the passage from universal principles to particular ethical dilemmas.

67 Paul Ricoeur, *Critique and Conviction* (New York: Columbia University Press, 1998), 92.

68 Congress, U.S., "Public Law 104–104," *Telecommunications Act* 8 (1996).

69 Nellie Bowles, "Thermostats, Locks and Lights: Digital Tools of Domestic Abuse," *The New York Times*, June 23, 2018, https://www.nytimes.com/2018/06/23/technology/smart-home-devices-domestic-abuse.html.

70 See, for instance, Sherry Turkle, *Alone Together: Why We Expect More from Technology and Less from Each Other* (New York: Basic Books, 2017).

71 Verbeek, "Materializing Morality," 361–80.

72 Don Ihde, *Technology and the Lifeworld: From Garden to Earth* (Bloomington: Indiana University Press, 1990), 144.

73 Ricoeur, *Oneself as Another*, 164.

74 Ricoeur, *Oneself as Another*, 273.

3

Ethical Deliberation of Digital Matters

A Deliberative Framework

This chapter explores how key elements in Ricoeur's narrative and ethical theories can be applied to the deliberation of the ethical issues that digital technologies raise. It describes the core characteristics of critical *phronesis*'s deliberation process and grafts onto this process a narrative approach that together can contribute to a plural deliberation that accounts for the peculiarities of completely new ethical dilemmas and integrates ethical principles, creative ethical imagination, diversity of deliberators, and contextual applications.

Although this framework is applicable in multiple deliberative contexts, from parents managing quotidian quarrels between children to judges weighing complex cases, I will keep our focus on ethical dilemmas involving Digital Poetics. Within this delimited scope, I propose three spheres of deliberation, whose descriptions and detailed analysis I offer in chapters 4, 5, and 6. I present them here briefly, so I can use them as testing scenarios for my proposals and arguments regarding narrative deliberation. These complementary dimensions point to the pervasiveness of digital technologies and the need for the deliberative process to take place in multiple social contexts and to be carried out by the creators of the technologies, their users, and the institutions that are transformed by them.

The first sphere of deliberation, based on the discussions of software engineering practice in the previous chapter, is one where software developers, designers, and managers meet to discuss the requirements of a new software functionality or application. Such meetings can follow a formal software development process or just be semi-structured meetings at start-ups. The second sphere of deliberation looks at users who constantly need to make choices about how, when, and for what purposes they will use new digital technologies. And finally, the third sphere of deliberation takes place when institutions need to create new ethical and legal standards, policies, and regulations for digital technologies, whether existing or under development.

This chapter is organized in five sections. In the first, I present the conceptual framework for grafting Ricoeur's narrative theory onto the phronetic deliberation process. The second section expands on the synchronous and asynchronous phronetic plurality concepts based on what Ricoeur describes as threefold *mimesis*. I argue that given the increasing complexity of digital technologies and their almost unlimited contexts of application, good deliberation necessarily involves the idea of a plural *phronimos* proposed by Ricoeur in *Oneself as Another*. The third section explores how cultivating narrative imagination fosters a more inclusive and fair deliberative process about new technologies. In the fourth section, I apply the concept of semantic innovation present in narrative configuration to the creative dimension of *phronesis*, a dimension that transcends rigid and often contradictory principles to find alternatives to ethical aporias. Finally, in the fifth section, I describe how the narrative approach has long-term implications and can be a source of new guidelines and principles for developing novel technologies through reflecting on poetic universals.

Phronetic Deliberation

In Praise of Phronesis

Since Aristotle, the concept of practical wisdom (*phronesis*) has been connected to the ethical deliberation process, and its primary outcome is good deliberation. This section highlights some key aspects of Ricoeur's analyses of *phronesis* that will be relevant to the next sections, in which I propose the integration of narrative theory with the phronetic deliberative process as a productive basis for the decision-making process of Digital Poetics.

In 1997, Ricoeur published an article with the suggestive title, "*À la gloire de la phronesis*,"[1] in which he analyzes the rehabilitation of the concept of *phronesis* by contemporary thinkers and discusses the hermeneutical difficulties arising from theoretical and practical environments as different as Aristotle's Greece and twentieth-century society. Ricoeur identifies two significant theoretical approaches in the efforts to update and revitalize the concept of practical wisdom from its ancient roots to current ethical discussions. The first is characterized by a certain interpretative freedom in relation to the classical sources. Ricoeur includes in this group the interpretations of Pierre Aubenque, Alasdair MacIntyre, and Hans-Georg Gadamer. The second group attempts to be very faithful to the text, which leads to detailed commentaries full of technical and specific observations, such as those by René Antoine Gauthier, Yves Jolif, William F. R. Hardie, and David Ross. Ricoeur, as usual, looks for a balanced and productive alternative between these two approaches, and he proposes a long path that starts from the investigation of the "labyrinth" of references and concepts in Book VI of the *Nicomachean Ethics* and gradually retrieves meaningful elements apt to enhance contemporary ethical debates through the application of practical wisdom.

The first essential aspect of practical wisdom that Ricoeur draws from the classical tradition and brings to current questions of human action and ethical deliberation is the relationship between the rational and desiderative (desire) functions of the human being. *Phronesis* is the cooperation of the desiderative part so that ethically good action can be carried out. Such dual allegiance of *phronesis* is what Aristotle called rational desire or desiderative rationality. Recognizing desire as a fundamental part of ethical action will be one of the foundations of Ricoeurian ethics, which starts from the assumption of the desire for a good life.[2] From the outset, this approach avoids an excessive rationalization of human action that neglects desire as an essential constituent of action.

Second, Ricoeur also asks himself about the epistemological status of practical wisdom. What kind of truth is associated with practical wisdom? In the classical conceptual framework, practical wisdom aims at something different from science, which seeks a form of general knowledge of necessary things. Aristotle defends a type of rationality and truth related to and appropriate for contingent things, a practical truth. But Ricoeur wonders about the epistemological status of such a practical truth. What, after all, is the truth of the contingent?

The striking features of practical wisdom lie in the shift from the rigor of a single, immutable answer to the realization of another degree of truth linked to the singular, the contingent, which seeks a fragile answer to the dilemmas of flesh-and-blood humans faced with ethical dilemmas created by unique and unprecedented circumstances. As Ricoeur summarizes: "The rich dictionary of phronesis invites us to explore a way of thinking that is human rather than divine, fragile rather than secure, singular rather than universal, balanced rather than excessive (no more mania!)."[3] In another text, Ricoeur offers a complement to this vision of truth in the domain of human action: "Few ideas are more salutary and liberating today than the idea that there is a practical reason, but not a science of practice. The domain of action is ontologically that of changing things, and epistemologically that of the plausible, in the sense of plausible and probable."[4]

Ricoeur strongly rejects any confusion and attempts to accommodate and apply the ideal of objective scientific truth to the contingent domain of human action. Practical wisdom has a human and humanizing characteristic in relation to the object it deals with (human things) and the way this type of rationality is constituted. It arises from attention to the opinions of people with experience. Experience in dealing with human dilemmas awakens and develops a type of intelligence that gives deliberators a privileged view that facilitates recognition of the singular situation in question and its relation to the respective ethical principles. Later on, this characteristic of *phronesis* will invite us to think about the importance of a plural debate, with participants from different backgrounds deliberating about digital technologies.

As a conclusion to his article on the glory of *phronesis*, Ricoeur proposes three reasons for the revival of the concept in contemporary thought that are pertinent to the conceptual framework of Digital Poetics. The first is *phronesis*'s special attention to the singular. Its epistemological kinship with sensation and equity in the realm of justice are essential sources of reflection for dealing with the impasses of the human situation that typically require situational judgment. While general principles and abstract laws provide a foundation for the deliberative process, there always remains an essential stage of consideration of the specificities of the ethical dilemma in question and an authentic translation of the general principles into a decision applied to the singular case.

A second reason for the resurgence of *phronesis* in contemporary ethics lies in its ability to reconcile universal principles with the ethos of the community in which it is required and exercised. According to Ricoeur, the revival of *phronesis* in contemporary times is due to this rooting of philosophical reflection in the existing practice of a living culture: "Philosophical reflection is rooted in a prior practice, in a living culture; in this respect, the exemplarity of the *phronimos* is emphasized, as is the link with the tragic *phronein* and political practice."[5]

Third, practical wisdom represents a way of confronting the demands for exclusivity of a type of rationality represented by the philosophical currents of Kantian formalism and Anglo-Saxon empiricism and by the mathematization of the human sciences. MacIntyre, Husserl, Heidegger, and Gadamer are representatives of the group of contemporary philosophers who approach phronesis as a form of resistance to the exclusivity of empirical rationality. Nevertheless, Ricoeur does not completely reject the demands of logical-theoretical rationality in the context of ethical deliberations. By contrast, the dialectic between Kantian universalism and practical wisdom is one of the backbones of Ricoeur's ethics that we shall explore next.

Practical Wisdom in Ricoeur's "Little Ethics"

In his "Little Ethics," Ricoeur consistently and creatively integrates practical wisdom with the logical tests of the Kantian categorical imperatives as a fundamental part of the arc of ethical deliberation. In the "Little Ethics'" conceptual arc, practical wisdom occupies particularly relevant places that define the theoretical architecture of Ricoeur's ethical proposal. In particular, practical wisdom is the foundation and starting point of the "Little Ethics" because its goal-oriented nature (teleological) prompts deliberation about a fulfilled life. Practical wisdom is also the pinnacle and point of arrival, the "final word" of the Ricoeurian "Little Ethics," as it reconciles universalistic deontological tests with the ideal of the good life in a judgment in a situation guided by solicitude and respect for others who are impacted in the deliberative process. The following is a decisive passage that synthesizes essential conceptual components for our study on the application of practical wisdom in Digital Poetics. In the following paragraph, Ricoeur, through a retrospective reflection, offers a synoptic view of the path that practical wisdom has fulfilled throughout the three studies of the "Little Ethics":

> Our final word in this "little ethics" that covers the seventh, eighth and ninth studies will be to suggest that the practical wisdom we are seeking aims at reconciling Aristotle's *phronēsis*, by way of Kant' *Moralität*, with Hegel's *Sittlichkeit.* Of *phronēsis* we retain the fact that its horizon is the "good life," its mediation deliberation, its actor the *phronimos*, and its place of application singular situations. But if at the end of these three studies the cycle appears to be complete, it is at another altitude, so to speak, that we now move, situated above our starting point: between the "naive" phronesis of our first pages (seventh study) and the "critical" phronesis of our final pages extends, first, the region of moral obligation, of duty (eighth study), wherein lies the demand that what ought not to be not be (namely, evil) and, more particularly, the demand that the suffering inflicted on humans by other human beings be abolished; and beyond this arid zone, extends the region of conflicts belonging to tragic action (ninth study). In this way, "critical" *phronēsis* tends, through these mediations, to be identified with *Sittlichkeit.* The latter, however, has been stripped of its pretention [*sic*] to mark the victory of Spirit over the contradictions that it itself provokes. Reduced to modesty, *Sittlichkeit* now joins *phronēsis* in moral judgment in situation. In return, because it has crossed through so many mediations and so many conflicts, the *phronēsis* of moral judgment in situation is saved from any temptation of anomie. It is through public debate, friendly discussion, and shared convictions that moral judgment in situation is formed.[6]

A detailed exegetical interpretation of this dense passage is beyond the scope of this section. However, I would like to highlight two ideas that are of particular interest for the following examination of narrative deliberation. First, there is an essential conceptual precision. Ricoeur distinguishes between naive and critical *phronesis*. In contrast to naive *phronesis*, critical *phronesis* has two essential elements. It is a kind of situational judgment that occurs after the deontological tests have already been applied to the deliberative process. In other words, critical *phronesis* emerges from the aporias created by the difficulty of applying moral principles and rules to unique contexts. Moreover, the core criterion for applying or adapting the general principle to the particular case is respect for the other human beings involved in the ethical decision. The intention is to stay as close as possible to the universal moral principle,[7] making the necessary adjustments to respect those involved in the particular case. Yet this decision should not be automatically "elevated" to a universal principle that would make the necessary exception to the singular case a general rule. Throughout this book, we use the concept of practical wisdom with the specific connotation of critical phronesis proposed by Ricoeur.

Second, the rapprochement that Ricoeur suggests between practical wisdom and Hegelian *Sittlichkeit* raises the fundamental problem of how practical wisdom can be applied in institutions. We are particularly interested in the importance of public debate and shared convictions, which foreshadow the forms of plural deliberation we will explore in the following sections and have crucial implications for the institutional sphere of deliberation in Digital Poetics.

Narrative Approach to Phronetic Deliberation

The Affinity between *Phronesis* and Narrative Structure

A narrative approach based on Ricoeur's narrative theory enhances the understanding of the deliberative process of practical wisdom. I suggest that to deliberate well is to create a concordant discordance[8] "aiming at a good life with and for others in just institutions."[9] This concordant discordance is, at once, composed of a dense retrospective narrative that considers as many events and connections as possible and, at the same time, is consistently articulated with a prospective narrative stemming from narrative imagination enhanced by fictional and historical stories, shaped by the respect of others, and guided by universal moral principles.

To delineate this concordant discordance, I return to the main characteristics of practical wisdom highlighted by Ricoeur in the previous section. For each, I will suggest possible contributions of a narrative approach to applying practical wisdom in the deliberative process. This analysis will be intentionally synoptic, and its primary goal is to project the broad outlines of narrative deliberation that I will explore more closely in the sections that follow.

Practical wisdom relates primarily to situational judgment that applies general principles to particular aporetic situations. A narrative approach offers an invitation to contextualize general principles through rich emplotments that foster more nuanced explorations of specific cases. Narratives consider facts, evidence, people, and technologies within social, political, economic, and historical contexts—what Ricoeur calls the conceptual network of action[10]—and integrate them in a synchronic order. Building on Ricoeur, I argue that in their fundamental structures, narratives integrate time (one thing after another), motivations (one thing because of the other), actions, events, people (characters), and physical and historical contexts (setting). Such a capacity to provide an articulated account of contextual situations is a requirement for good deliberation.

I offer five traits of Ricoeur's theory of narrative that correlate with and deepen the nature of practical wisdom. As will be evident, while at moments Ricoeur himself offers these identifications, at other points I draw comparisons on the basis of his narrative theory that he himself does not. First, there is an affinity between narrative understanding and practical wisdom. Ricoeur recognized this affinity in multiple texts.[11] In "Life in Quest of Narrative,"[12] commenting on Aristotle's *Poetics*, Ricoeur asserts that narratives foster "a sort of understanding that can be termed narrative understanding and which is much closer to the practical wisdom of moral judgment than to science."[13] In *Time and Narrative*, Ricoeur explores this affinity through the connection between practical understanding and narrative understanding: "What then is the relation of our narrative understanding to this practical understanding? The answer to this question governs the relationship that can be established between the theory of narrative and that of action."[14] He goes on to explain that the theory of narrative actualizes and integrates actions taken separately into a syntagmatic unit over time: "In passing from the paradigmatic order of action to the syntagmatic order of narrative, the terms of the semantics of action acquire integration and actuality."[15] We see this movement from the paradigmatic order of the semantics of action to the syntagmatic order of narrative understanding at the core of deliberation that moves from principles and paradigmatic actions to particular situations in which actions and events integrate and actualize in time.

The second core characteristic of Ricoeur's critical practical wisdom that can be enhanced through a narrative lens arises in its plurality of opinions, reasons, and interpretations within the public debate. Narratives offer a privileged way of integrating, through the interweaving of narrative plots and narrative lines, the plurivocality of phronetic deliberation.

Third, practical wisdom's process of deliberation offers the integration of past events into a decision about how the future should develop to restore the possibility of a good life. Narratives are always at a certain point in time—*in media res*—contemplating with a double look—like the two-faced Roman deity Janus—the past and the future, so at each point in the narrative, it is possible, or should be, to think about why we are here and where we want to go.

Fourth, practical wisdom deliberation is teleological; it aims at a state of affairs that maximizes the possibility of a good life. Narratives are structurally teleological as they organize events and accidents in a temporal fashion that leads to a conclusion. In "Narrative Time," Ricoeur says that "there is no story if our attention is not moved along by a thousand contingencies. This is why a story has to be followed to its conclusion."[16] The teleological structure of narratives renders them understandable, and such a goal-oriented structure provides a responsive fit for the exercise of practical wisdom.

Finally, practical wisdom does not consist of an automatic application of general principles to particular situations, as it is inventive and proposes new decisions for unprecedented ethical dilemmas. It guides deliberators to the solution of ethical aporias when, through examining the details of the particular situation, it can invent a new course of action leading to a good life with and for others. Such an innovative dimension finds its counterpoint in another feature of narrative that is tied to creative imagination and engenders plots emerging from disjointed events that build to a concordant discordance.

For each of Ricoeur's critical practical wisdom core dimensions—contextuality, practical understanding of action, temporality, creativity, teleology, and plurivocality—we find features of Ricoeur's narrative theory that illuminate the nature of the phronetic deliberative process and enhance its outcome. Narrative imagination offers, then, an essential tool for good deliberation, particularly, as we shall see, in cases involving new digital technologies that stretch existing principles, norms, and decision precedents to the point of rupture.

Although Ricoeur's work highlights the importance of narrativity to understand the human experience, the fundamental role of practical wisdom in dealing with ethical dilemmas, and even the relationship between ethics and narrativity, it is intriguing that he does not develop a deeper analysis of the possibilities that his narrative theory offers for thinking about the process of phronetic deliberation. One hypothesis for Ricoeur's silence, which surfaced in his debate with Peter Kemp, is his hesitation regarding the idea of narrative ethics.[17] Fundamentally, Ricoeur seemed concerned with a narrative ethics that is not anchored in moral principles. I am well aware of this risk as I defend the idea that any narrative framework seeking to encompass Ricoeur's take on critical practical wisdom must incorporate all three moments of his "Little Ethics," including the Kantian universalization criteria. The deontological test can be thought of as a set of limits on what the prospective narrative should not and cannot narrate, that is, the evil and violence that destroy the possibility of others to be. Hence, in the construction of deliberation, narrative imagination and principles of respect for the other cooperate in a dialectical way, shaping the prospective narrative. Deontological principles impose limits on the content that should be expressed through narrative deliberation. Even in cases where care and respect for those impacted by the ethical decision require adjustments to the general rules, those rules can and should continue to sway the deliberative narrative.[18]

It is also critical to note that I am not proposing exclusively a narrative approach to the deliberative process. Just as fictional or historical narratives have descriptive segments, formal argumentations, and simple lists of things or facts integrated into the narrative plot, the deliberative process benefits from a narrative approach even though it has non-narrative components, such as the description of technical artifacts or the discussion of legal structures and procedures. I am therefore not claiming the exclusivity of the narrative approach in phronetic deliberation but rather its critical importance to the overarching deliberative process.

Narrative-Based Deliberation

On the basis of the interrelation between narrative and phronetic wisdom just offered, I suggest a model in which we assume that an ethical decision always takes place at the end of a deliberation that is based on a phronetic narrative that intends a desired ethical situation in the future. Phronetic narratives are the results of deliberations of practical wisdom that are shaped by narrative structures. Phronetic narratives have two main components. Retrospective narratives interpret actions, motives, and events that took place before the time of deliberation in an effort to understand the state of affairs at the time of deliberation. Prospective narratives both extend retrospective narratives to create a phronetic narrative that leads to the ideal of a common good life and also creatively propose a set of actions that enhance the likelihood of reaching a state of affairs we want to inhabit. Let's look at each of these moments more carefully.

Retrospective Narratives

Retrospective narratives within ethical deliberation are close to historical narratives in that they aim at the truth of what has happened. Richard Kearney explains that the difference between fictional and historical narratives is based on the fact that

> in historical narratives, clearly, we see the gap between narration and action narrowing to the degree that the telling now unfolds in the realm of the "as": historians try to tell the past "as" it happened, not "as if" it happened. In other words, even though in history the "as" (*wie*) always retains a certain "figurability"—we never have a completely literal account—it is much closer to the real world in that it must observe certain "truth claims" which the "as if" mode suspends.[19]

The retrospective narrative seeks to figure out past actions and motives in order to come as close as possible to what actually happened. Kearney speaks of a "trans-narrative truth"[20] that emerges from the plurality of narrative interpretations and which has as its fundamental criteria the relationship with other narratives and a real relationship with the facts and events.

Like historical narratives, retrospective narratives aim at "the operation that consists in providing ourselves with a figure of what was."[21] This process of the historical imagination, which Ricoeur explores in *Time and Narrative 3* and *Memory, History, Forgetting*, is fundamental to the retrospective deliberative process. It also includes an aspect that transcends the explanation of facts on the cognitive plane but makes us—through the narrative approach—"at the same time see, feel and live the way things actually happened."[22] Retrospective narrative finally joins ethical deliberation through what Ricoeur calls a "debt to the past,"[23] through which Ricoeur emphasizes the ethical task of historians to explore what happened through documents and testimony-based historical imagination so that the truth of the past is recognized and evil and violence are not repeated. This task is similar to that of phronetic deliberation in taking a retrospective narrative look at the events and actions that created an ethical dilemma.

In his introduction to narrative theory, Porter Abbott discusses the application of narratives in judicial contexts through the famous case of the American athlete O. J. Simpson: "The Simpson trial, like any trial, was a contest of narratives involving two parties, each seeking to bring the facts (or what could be adduced as facts) into conformity with a coherent narrative that favored its conception of the accused."[24] While Porter explores this case in connection with the influence of masterplots that create biases about narrative judgments—critical for narrative deliberation as well—he emphasizes that all trials are indeed narrative contests.

Ricoeur's famous hermeneutic principle of the conflict of interpretations captures this contest more positively and productively. For instance, discussing H. L. A. Hart's concept of juridical judgment, Ricoeur says that "the intermediary function of juridical reasoning clearly shows that the procedures of validation have a polemical character. In front of the court, the plurivocity common to texts and to actions is exhibited in the form of a conflict of interpretations, and the final interpretation appears as a verdict to which it is possible to make appeal."[25] Within our narrative framework, each proposed retrospective narrative thread considered within the phronetic deliberation represents the plurivocity that a judgment based on practical wisdom must address.

Nevertheless, unlike a typical judicial trial, phronetic deliberation is not just the process of determining a single winning narrative. While in some cases it may be possible, as Kearney suggests, that the historical inquiry is able to sort out the competing retrospective narratives into one that is most probable, in other cases, it may be possible to construct a meta-narrative—a concordant discordance—out of the conflicting narratives. The potential meta-narrative need not create unanimous agreement about what happened or solve all conflicts, so it may continue to express tensions inherent to healthy clashes of interpretations. Ricoeur's quote also alerts us that the creation of these phronetic narratives must always be open to "appeals." Phronetic deliberations are constantly open to revisions, extensions, and reconsiderations. This is another affinity with narratives, which are naturally elastic in the way they express our temporal experiences. It may be possible to tell the narrative differently or tell it more.

As we will see later, such dynamics are particularly relevant in decisions about digital technologies that are often challenged by the digital plasticity and pervasiveness we discussed in the first chapter. For example, social media application developers and users must integrate into their narrative deliberation the various stories of those users, particularly young people and children, who are negatively impacted by the excessive use of these types of technologies. Institutional deliberators should also integrate into their deliberations on policies for adopting and regulating new technologies based on artificial intelligence the various stories about biases inherent in the training data used to generate models of these types of technologies.

When applied to digital technologies, retrospective narratives should weave together the history of technology enablers, users' test reports run with these enablers or similar artifacts, and technical specifications that would constrain prospective deliberation. The narrative contest is not focused exclusively on absolving or condemning a certain digital technology as a whole.[26] Rather, it explores aspects, attributes, functionalities, and applications of digital technologies that could be harmful or counterproductive to a flourishing and good life or, conversely, productive of those ends. This multifaceted, contextualized, and nuanced analysis is a fascinating aspect of narrative deliberation as it explores combinations of both retrospective and prospective narratives.

Prospective Narratives

A phronetic decision does not simply try to address retrospective narratives, for it also has a prospective dimension as well. It should encompass a prospective narrative attentive to the goals lying ahead. In my view, the phronetic decision should aim to create a concordant discordant narrative, bringing to bear retrospective and prospective narratives that are bound by three complementary and equally important dimensions: the universal respect for the limits moral principles impose, the actual description of facts and actions, and the possible ways digital technologies will be developed and used. Here, I want to focus on a phronetic decision's prospective factors and the prospective narrative created.

A decision itself can be envisaged as a post-deliberation narrative, that is, a prospective narrative. The process of phronetic deliberation is, therefore, the weaving of an encompassing narrative that mediates the tensions of retrospective narratives with a concordant discordant prospective narrative. This overarching emplotment is what I call "deliberative narrative." In decisions about digital technologies, the digital materiality discussed in the first chapter suggests that various aspects of the decision can be encoded in the form of algorithms, user interface aspects, configuration parameters, input data selection criteria, default values of data fields, and so on.

Retrospective narratives provide bases for, and also limit, prospective narratives horizontally (syntagmatically). It is impossible to maintain the narrative plot if the retrospective segment is not properly integrated into the prospective segment. Similarly, my proposal is that moral principles, respect for others, and the desire for a good life with and for others also provide the bases for and limit the prospective narrative vertically (paradigmatically). They should impose a limit on the prospective narratives of phronetic deliberation.

To instantiate and clarify these concepts through a generic example, when a judge sentences someone to prison after a fair trial, ideally, two complementary aspects should be considered. First, the decision is the result of a dense retrospective narrative that accounts for as many narrative threads as possible about the facts and actions leading to the judgment. Second, the prospective narrative should involve education and social reintegration and not just a narrative of harmful experiences in prison environments. Finally, this decision, constrained vertically by the categorical imperative, blocks narratives where the convict is reduced to the character in a prospective narrative who disrespects the rights of other people in society.

As we shall describe in more detail later, the exploration of possible narratives involving digital technologies operates as a test laboratory for prospective narratives involving digital artifacts. It is important to acknowledge that, in addition to narrative elements, there might also be other descriptive outcomes of the deliberation. These outcomes will prescribe attributes, characteristics, and technical limits of digital technologies that maximize their probability of being part of the proposed prospective narratives and limit other possible narratives. Finally, the deliberative narrative has unique characteristics given its application to ethical judgment. It integrates a component of historical narrative, which seeks to approach what actually happened, with an element that has yet to happen, that is, with a possible and desirable future closer to the possibilities of existence that fictional narratives open.

At various points in his work, Ricoeur discusses the responsibility of historical narratives to recount the suffering of victims to ensure their experiences are remembered and to prevent the recurrence of violence. For example, in *Time and Narrative*, he affirms: "We tell stories because in the last analysis human lives need and merit being narrated. This remark takes on its full force when we refer to the necessity to save the history of the defeated and the lost. The whole history of suffering cries out for vengeance and calls for narrative."[27] In *Memory, History, Forgetting*, he expresses the possible healing properties of historical narratives, saying that "the historian, providing a place for the dead, makes a place for the living."[28] And in *Oneself as Another*, Ricoeur suggests the intrinsic relationship between narratives and desirable futures: "moral judgment needs the art of storytelling to schematize its aims."[29] The schematizing capacity of narratives brings to phronetic deliberation the responsibility toward the victims of the past and the hope that prospective narrative creates a better form of common life for those alive now and in the future.

The meaning-structuring capacity of narratives also allows ethical deliberation to take into account a hierarchy of goals on variable time scales. Ricoeur comments, for example, on how MacIntyre uses narrative structure to propose a hierarchical structure of human action that can be analyzed not only on the basis of various practices (as we talked about in software engineering) but also by means of a higher-level meaningful structure such as life plans and then finally the whole extent of human existence through the concept of the narrative unity of life.[30] This flexibility of narratives is productive for deliberating on the varying scopes of the consequences of certain choices. Particularly in the case of new technologies, it is common to think of the benefits of short-term prospective narratives but also to neglect the analysis of long-term arcs, as in the case of the environmental impacts of certain technological innovations.

Ricoeur's narrative theory also sheds light on another fruitful dimension of the "openness" of prospective narratives. Narratives propose a possible world in which those involved are invited (in some cases compelled) to participate. In the case of digital technologies particularly, users always reappropriate the prospective narratives that shape artifacts and can reinterpret them in ways unforeseen at the time of deliberation. In this sense, the phronetic deliberation process must be understood as a hermeneutic spiral that is always dynamic and constantly open to creative reinterpretations. The personalization of digital technologies brings new possibilities for this hermeneutic spiral to involve a constant reassessment of the prospective narratives of a given artifact based on how users appropriate it. This will be a topic of further analysis in the users' sphere of deliberation.

Phronetic Narrative

The phronetic narrative, then, is a complex emplotment weaving together retrospective narratives that capture what happened with a prospective narrative that proposes a possible future that can be narrated on the basis of certain actions and choices made.[31] I want to highlight here something that has been thus far only a subtext: the *ethical* dimension of the phronetic narrative and its consequences. In *Oneself as Another*, Ricoeur discusses the role of narratives as mediators between a theory of action and an ethical theory. He writes, "[N]arrative theory can genuinely mediate between description and prescription only if the broadening of the practical field and the anticipation of ethical considerations are implied in the very structure of the act of narrating."[32] This idea of mediating between description and prescription is fundamental enough to play a foundational role in the structure of the studies in *Oneself as Another*. Yet while Ricoeur in that context is primarily concerned with how narratives add a semantic layer on top of isolated actions leading to his concept of narrative identity, phronetic deliberation makes it clear that narratives locate the dialectics of description and prescription within the core of ethical deliberation, inviting us to consider additional aspects of our narrative approach to critical practical wisdom.

The critical practical wisdom framework proposed by Ricoeur's "Little Ethics" imposes certain conceptual boundaries on the phronetic narrative. First, a deontological test must guide the deliberative narrative and its prescriptive dimension; therefore, it cannot narrate just anything. The evaluation of prospective narratives must always be based on the criterion of respect for the dignity of all people involved in the decision. Second, the deliberative narrative must create a consistent and achievable emplotment that coherently concatenates the retrospective and prospective narratives. The discordant concordance characteristic to which every narrative aspires is fundamental for the consistency of the deliberative narrative that mediates description and prescription. Third, the deliberative narrative proposes new possibilities of temporally configured actions to overcome the existing ethical dilemma. Therefore, the deliberative narrative needs to leverage its intrinsic creative narrative imagination to find new ways to apply moral principles to particular aporetic situations. The prescriptive pole of the phronetic narrative is anchored by its descriptive pole but not given by it. Fourth, the deliberative narrative must incorporate as many and as nuanced perspectives as possible in its descriptive dimension so that its prescriptions are fair and consistent with the ideal of a fulfilled life within just institutions. I propose considering the descriptive potential of phronetic narrative in terms of its narrative density.

Narrative Density

In the *Statesman*, Plato compares a good legislator with a skilled weaver. Their common defining capacity is the ability to intertwine different materials into a cohesive and high-quality final product.[33] While the weaver chooses to select threads and skillfully uses the loom to produce good fabric, the legislator organizes a *polis* (political body) based on the diverse qualities of its citizens. Although Plato's objective and arguments are not particularly relevant here, his metaphor depicts at least two essential aspects of narrative density. Examining the legislator's role ties quite effectively to our analysis of good deliberation while visualizing narratives as interwoven threads illuminate the plot's deliberative threads.

Just as the total number of threads per square inch indicates fabric quality, durability, and tightness, the number of threads considered in a phronetic narrative is an indicator of deliberative depth. However, this quantitative metric alone is not enough, as it has to be qualified by the diversity of threads capturing different perspectives on the ethical dilemma. These perspectives should include minorities, marginalized, and suppressed voices in particular. Including these complementary views provides a multifaceted yet cohesive decision fabric, and the coherence of the emplotment weaves a tensional concordant narrative from the potentially discordant perspectives. This diverse and cohesive thread count of the narrative deliberation comprises its narrative density.

The two segments of the phronetic narrative can be evaluated in terms of narrative density. Both retrospectively and prospectively, the threads' multiplicity manifests in the conflicting and complementary descriptions of facts and actions that are assessed, selected, and integrated, and in the diversity of voices and perspectives that explore different narrative contexts and consequences. The density of a deliberative narrative is, therefore, the set of different narrative threads that represent diverse configurations and combinations of retrospective and prospective narratives.

The retrospective narrative density is the sum of voices and points of view integrated into the narrative. The greater the consideration and coherent correlation of these views, the denser the retrospective narrative tends to be. In this sense, the participation of multiple representatives with different skills, perspectives, and values makes the narrative denser. For instance, when deliberating the new functionality of a social networking artifact, not only managerial and technical voices must be heard, but also participants with a background in public policy, social psychology, pedagogy, social science, digital privacy, and security, to name only a few.

The richness of the prospective narrative also defines the narrative density. Although the prospective narrative in the decision may represent only one or a few threads of action and events suggested, all possible prospective narratives considered in the deliberative process contribute to the density of creative imaginative variations that lead to a decision. It is specifically at this point that imaginative variations and fictional narratives play an essential role in configuring deliberative narratives with higher density, which is another theoretical and practical advantage of integrating narrative theory into the analysis of phronetic deliberation. I explore this aspect further in the final part of this chapter.

After suggesting a network of concepts and arguments supporting narrative deliberation, I would like to conclude this section by taking stock of the conceptual framework proposed so far. Starting from the idea of infusing Ricoeur's narrative theory into critical practical wisdom, we have proposed the concept of phronetic narrative that re-schematizes the deliberative process by weaving together a retrospective narrative—akin to historical narratives as it focuses on facts and actions that led to the ethical dilemma—and a prospective narrative—that creatively proposes a possible world as a set of actions that aim toward the realization of moral principles while considering the contextual respect of each person involved in the ethical deliberation.

We then qualified the phronetic narrative through the concept of narrative density. This analysis lays the foundation for applying other dimensions of the narrative framework to the core characteristics of Ricoeur's critical practical wisdom. These examples include the threefold *mimesis*, narrative imagination, and the poetic universal. The exploration of these complementary dimensions of narrative deliberation constitutes the next objective in this chapter, starting with the potential of the threefold *mimesis* to deepen our understanding of plural phronesis.

Plural Phronesis

The topic of narrative density reminds us of the nature of *phronesis* as plural, discussed in the prior chapter. In this second section, I want to develop the implications of plural *phronesis* for narrative deliberation. The fundamental insight of plural *phronesis* as considered previously in chapter 2 is that as the deliberative process captures more voices, more possibilities for comprehensive, inclusive, and fair decisions will emerge. A narrative approach goes further by breaking down the concept of plural phronesis into synchronous and asynchronous dimensions. This finer-grained analysis suggests practical ways that a narrative approach plays a crucial role in enhancing phronetic plurality, leading, therefore, to better deliberations.

Synthesis of the Heterogenous and the Synchronous Dimension

The synchronous dimension is the most evident and is related to the diversity of participants in the deliberative process. As mentioned in the previous chapter, in *Oneself as Another*, Ricoeur provocatively asks whether the *phronimos* needs to be just one person. He responds negatively to this rhetorical question by privileging public debate and its potential for overcoming ethical aporias. The narrative structure allows the continuous complexification of descriptions and the exploration of temporal and causal relationships, making it ideal for synthesizing different opinions and perspectives of the synchronic plurality.

In the sixth study of *Oneself as Another*, Ricoeur examines how narrative acts as a synthesis of the heterogeneous. Every narrative composition is a discordant concordance that performs a synthesis of heterogeneous aspects mediated by narrative emplotment.[34] Among these mediations, Ricoeur highlights the mediation between the manifold of events and the temporal unit; the mediation between disparate components (intentions, causes, and chance occurrences) and the sequence of the story; and the mediation between the pure descriptive succession and the unity of temporal form. This narrative capacity for synthesizing the heterogeneousness that occurs through emplotment is fundamental for phronetic deliberation and its respective decision to emerge from the debate in which multiple participants offer different perspectives for evaluating the ethical problem in question.

Threefold Mimesis and the Asynchronous Dimension

The contribution of the narrative approach to plural *phronesis* becomes even more apparent when considered in light of the asynchronous dimension of deliberation. Each time deliberators come into contact with a historical or fictional narrative, particularly those involving the specific object of deliberation, they refigure their symbolic universe and their horizon of meaning, including new possibilities for seeing their object of deliberation from new perspectives. As I shall pursue, this process is intrinsically linked to Ricoeur's hermeneutic movement of threefold *mimesis*[35] and suggests that deliberators will be able to consider alternatives and create new possibilities from the richness of their symbolic universe engendered by collecting multiple and diverse narrative figurations.

The concept of threefold *mimesis* proposes an asynchronous, bidirectional, and iterative relationship between narrative and reality. It unfolds in three interrelated moments. In the first moment, *prefiguration* or *mimesis* 1, there is a focus on our capacity and prior knowledge that allows us to understand and interpret events. Even before a story is told, we already have a set of expectations and understandings about the world that prepare us to make sense of what will be narrated. The second moment, *configuration* or *mimesis* 2, deals with the creation and structuring of the narrative. Here, events are selected, organized, and connected into a plot. In this process, the narrative gives shape and meaning to the events, proposing a coherent interpretation that may not be evident in the disorganized reality. Finally, the moment of *refiguration* or *mimesis* 3 reflects the impact of the narrative after it has been told and how it reshapes our understanding of the world. This stage highlights the narrative's ability to transform our view of reality, influencing our perceptions, beliefs, future actions and, in our particular case, ethical deliberations.

The threefold *mimesis* offers a perspective on how stories not only reflect reality but also actively shape it, influencing our experience and understanding of the world. Hence, the symbolic universes that deliberators bring to the deliberative process result from an ethical-narrative formation process. Historical, fictional, and journalistic narratives are ethical laboratories, as Ricoeur suggests,[36] and the result of these experiments in narrative configurations leaves an ethical residue that expands the deliberators' horizon of meaning. For instance, each time deliberators engage with a narrative related to a digital artifact enabler, these narratives reshape their understanding of the artifact, expanding it based on the narrative threads that fictional and historical narratives propose. By configuring social, political, environmental, and economic contexts that digital technologies impact, narratives offer possible elements for prospective narratives that can be accepted or explicitly rejected during the deliberative process.

In addition to new configurations that deliberators gain about possible contexts and consequences of new technologies through contact with fictional narratives, extensive contact with the narrative form itself also influences the process. As will be discussed in greater detail in the following sections, a narrative understanding enhances the engagement with conceptual action networks fundamental to nuanced and deeper deliberation. Deliberators with a narrative understanding polished by extensive contact with fictional and historical narratives find it easier to recognize the possible variations of the plot, the settings in which the technology plays its role, potential interactions between characters, and the complex ways cause and effect can manifest in plot development.

Thus, Ricoeur's threefold *mimesis*, which puts the narrative in dialectical contact with the lived world and emphasizes the refigurative impact on readers, provides a hermeneutical key to expanding the concept of plural *phronesis*. Phronetic narratives are ideal for accommodating diverse synchronic voices of public debate, and the epistemological implications of narrative refiguration open up a new asynchronous dimension to plural *phronesis*.

Turning Aporias into Euporias with Narrative Imagination

The next feature of critical practical wisdom explored within the narrative framework, the topic of section 3, is a creative solution to aporetic moral dilemmas when approached by a deontological method of simple instantiation of universal principles. Faced with the practical paralysis of the simple logical derivation of moral postulates that often fail to account for the complexity of unprecedented ethical situations, practical wisdom is a creative force that invents a contextual decision based on universal principles focusing on respect for others. It is precisely in this operation that *narrative imagination* makes a substantial contribution to the understanding of critical practical wisdom.

To investigate the contribution of narrative imagination to phronetic deliberation, I propose, building on Ricoeur, a conceptual trajectory that begins with the concept of imaginative variations that Ricoeur explores based on Husserl's phenomenology and advances toward the idea of creative imagination that extends Kantian epistemological theory. This analysis allows Ricoeur to expand from the descriptive nature of the Husserlian concept to the more open and creative possibilities of the Kantian concept. Finally, I explore in detail how creative imagination takes place in fictional narratives, exploring new possibilities of being in the world alongside digital innovations.

Imaginative Variations

In the context of Husserlian phenomenology, imaginative variations are a fundamental step in the process of what is called eidetic reduction. After identifying a phenomenon to be described, the phenomenologist proposes variations concerning the attributes and characteristics of a given instance of the phenomenon, searching for invariant structures. Thus, "if the circle with unbroken line is the exemplary original and the broken-line circle's variants upon it, the shaded area at the center is the common core with regard to which they exhibit a congruence: It is the set of features constituting the eidos of the class under consideration."[37] Mohanty explains that Husserl uses the method of imaginative variations to explore the essence of regions of phenomena, whether materials, works of art, moral experience, or the experience of consciousness.[38] Ricoeur also emphasizes the conceptual importance of imaginative variations as a fundamental step in Husserl's phenomenology: "The role of imaginative variation in the Husserlian reduction from fact to essence cannot be overestimated."[39]

For phenomenology, the actual existence of the proposed variations is suspended in phenomenological reduction. Imaginative variation is, therefore, an exploration of the essential possibilities of a given phenomenon and not an empirical exploration of its characteristics. Ricoeur stresses and values this non-reductionist position in the empirical domain and states that "phenomenology is also a victory over brute fact by the method of imaginative variation."[40] I adopt this non-empirically constrained approach as we move toward narrative imagination. By contrast, in Husserl's phenomenology, the concern is clearly with the description of the essences the real phenomena give, whether external or of consciousness. Husserl emphasizes: "The decisive factor lies before all in the absolutely faithful description of which lies before one in phenomenological purity, and in keeping at a distance all interpretations that transcend the given."[41]

Husserl is primarily concerned with an existing essence, whereas my approach moves beyond the descriptive limits and ontological presuppositions of the Husserlian method. Like Ricoeur's own trajectory, my work advances toward a hermeneutic understanding of phenomena. Along the same lines, Mohanty suggests the application of imaginative variations to clarify meanings and not essences in the phenomenological sense. The application of imaginative variations as a method of exploring meanings and not simply describing essences makes it more suitable to our narrative framework, avoids metaphysical discussions not relevant to the central focus of our research, and maintains its heuristic potential as a way of describing and clarifying the meanings of digital technologies.

Without explicitly exploring the dichotomy between phenomenological essence and hermeneutical meaning, Moustakas helpfully employs the term imaginative variation to concisely and productively clarify its role:

> The task of Imaginative Variation is to seek possible meanings through the utilization of imagination, varying the frames of reference, employing polarities and reversals, and approaching the phenomenon from divergent perspectives, different positions, roles, or functions. The aim is to arrive at structural descriptions of an experience, the underlying and precipitating factors that account for what is being experienced.[42]

Moustakas's definition is particularly interesting for our specific argumentation because it includes fundamental elements for considering imaginative variations expressed in the forms of narratives. The different frames of reference in narratives are the different contexts in which digital technology is placed. Exploring these new frames of reference reveals other existing or possible aspects of technology enablers. Polarities and reversals appear in fictional narratives as surprises in plots that Aristotle's *Poetics*[43] already highlighted in Greek tragedies. The various combinations of roles, perspectives, and consequences of using digital artifacts depicted by fictional narratives find a direct counterpart in Moustakas's definition of imaginative variations within the phenomenological method. Nevertheless, Moustakas's conceptual framework is still within a restrictive contour that emphasizes the existing state of affairs rather than those possible.

Productive Imagination

Given the understanding of imaginative variations, we can advance the inquiry into Ricoeur's creative appropriation of the concept and eventually consider the contributions of narrative imagination to critical practical wisdom. An example of Ricoeur's employment of imaginative variation, drawn from his early work *Fallible Man*, will be useful:

> No doubt the understanding of the primordial first, then of the fallen in and through the primordial, requires a kind of imagination, the imagination of innocence or a "kingdom" wherein the quests for "having, power, and

> worth" would not be what they in fact are. But this imagination is not a fanciful dream; it is an "imaginative variation," to use a Husserlian term, which manifests the essence by breaking the prestige of the fact. In imagining another state of affairs or another kingdom, I perceive the possible, and in the possible, the essential.[44]

Imaginative variation is not a "fanciful dream," because it employs existing examples to present a version of what is possible. It is possible to conceive of a kingdom—that is still a kingdom—without the "quest for having, power, and worth." While the passage from the instance or exemplar to its underlying essential structure remains central, Ricoeur's primary interest in imaginative variations lies in the exploration of the possible. This exploration is a critical function of imaginative variations that break free from the real exemplar and explore its possible variations, new contexts, and what is unveiled through the use of imaginative variations that stretch the boundaries of perception through imagination. In the case of digital technologies, breaking the "prestige of the fact" might entail disrupting the prestige of the main use cases typically publicized as the principal features and exploring what else that technology is, or could become. A Bluetooth chip tile may track your lost backpack, yet it may also become a tracking mechanism that compromises your privacy.

Ultimately, Ricoeur finds Husserl's approach toward imaginative variation too limiting because it is too focused on an essence. In his mature works, Ricoeur shifts his focus from a comprehensive description of the phenomenon as it appears toward humans' imaginative capacity to explore what is not yet here. While Husserl privileges the description of essences, Ricoeur changes his focus over time to emphasize the exploration of the possible in the creative imagination.[45]

Understanding Ricoeur's shift in focus requires revisiting his exploration of imagination through Kantian lenses, specifically the contrast between productive and reproductive imagination. While the reproductive imagination is linked to the schematic organization of empirical sensibilities and therefore anchored in the experiences of phenomena, the productive imagination creates new images and constructs possible schemas that are still unrealized.

For Ricoeur, Husserl's use of imaginative variations is primarily a form of reproductive imagination as it explores new schemas to understand existing phenomena. It aims to describe the essence of what is experienced. But Ricoeur thinks imagination has the potential to go beyond the description of what is and propose something new, something that may possibly become real but is not yet actual. Just as Ricoeur surpasses Husserl, here he surpasses Kant. For Kant, the productive imagination does not extend beyond the realm of aesthetics, while for Ricoeur the productive imagination has implications for who we are.

Chown and I[46] have developed how metaphors are forms of productive imagination used in the technology design process. The ability to "see as" is necessary for metaphorical thinking and essential for ethical deliberations. For instance, viewing a user's data as "digital gold" could shift a designer's approach to privacy and security, emphasizing protection and value. Along these lines, Donald Schön,[47] for example, studied how metaphors can frame both design processes and public policy in different ways. In public policy, viewing a social issue as a "war" or a "journey" can lead to vastly different approaches and solutions. Similarly, in technology design, metaphors can determine whether a product is developed with a focus on user empowerment or user dependency.

The relevance of creative imagination to ethical deliberation is also suggested when Ricoeur explores imagination in the context of social imaginaries:

> [W]e are not able to take hold directly of the creative dimension of imagination. We become aware of it thanks to the interplay between ideology and utopia. And that means that we do not even directly grasp the genuine function of ideology as the power to provide us with an image of what we are, nor the originary function of utopia as the power of otherness.[48]

Both ideology and utopia are forms of creative imagination. Ideology, in its positive manifestation, provides a creative image of a social group's identity, and utopias propose ways of being that challenge fixed structures and sedimented relationships as a social group.

Ricoeur highlights other key features of creative imagination when he applies the concept to analyze utopias: "What particularly interests me is the notion of utopia is that it is an imaginative variation on power. . . . The result of reading a utopia is that it puts into question what presently exists; it makes the current world seem strange."[49] Utopias can foster creative imaginations of power that allow us to see new possibilities beyond what exists. Utopias allow us to see the real as strange, and they open up the possibility for reaction to reality or creation of the possible.

Utopian narratives point to Ricoeur's appropriation of imaginative variations as potential forms of productive imagination, contrary to Husserl, and the broader relationship between imaginative variations and fictional narratives in Ricoeur's work. "Literary fiction is an imaginative variation whose premises the reader assumes for a while."[50] In *Time and Narrative*, the creative potential of the imagination applied to narrative emplotment emerges in full force and structures fundamental aspects of the work, such as the concept of configuration. "We ought not to hesitate in comparing the production of the configurational act to the work of the productive imagination."[51] Fictional narratives thus may become privileged ways of exploring and representing productive imagination structured in time. Through their plots, narratives explore alternatives to the existing reality. I explore next the figurative and shattering potential of fictional narratives, considering them in the context of phronetic deliberation. This discussion amplifies the horizon of imagination from the already fruitful consideration of variations of existing realities—akin to the efforts of retrospective narrative—to the exploration and proposal of alternative and possible realities, similar to the primary task of prospective narratives along the arc of phronetic narratives.

The Refigurative Potential of Narrative Imagination

Each plot schema that fictional narratives suggest can be integrated into the consideration of the best prospective narrative to expand the deliberative reach toward as many scenarios as possible and to account for the multiplicity of ways technologies will affect others. Enlarging the horizon of meaning on the object of deliberation is a particularly important way to respect those whom the phronetic decision will affect. Hannah Arendt's concept of representative thinking within her theory of political thought captures this propaedeutic role of narratives through various standpoints deliberators encompass for their ethical decision: "The more people's standpoints, and the better I can imagine how I would think and feel if I were in their place, the stronger will be my capacity for representative thinking."[52]

The representative thinking emerging from narrative refiguration will never be equivalent to a synchronic plurality of different voices participating directly in a deliberation. Nevertheless, it might play two essential roles in the deliberative setting. First, it may be the best assurance of voicing the perspectives of minorities and marginalized people who cannot directly participate in the deliberative process, though representative thinking should never replace the possibility of direct participation of all possible representatives. Second, representative thinking might develop an ethical sensibility to others' arguments and viewpoints by creating a shared narrative framework that captures in different ways the object of deliberation. For instance, it can foster a disposition to explore diverse perspectives on how a mobile application deployed worldwide will impact users' lives. This shared perspective might be critical in the deliberative process as it paves the way for broader perspectives and openness to external arguments and preoccupations.

As already suggested, this augmentation of deliberative narrative density can be approached through Ricoeur's hermeneutic analysis of the triple *mimesis*. Each cycle from prefiguration to refiguration through a configured narrative enhances the deliberator's hermeneutical horizon. The deliberator expands his or her horizon of possible meanings through a new refiguration triggered by fictional and historical narratives. Such an enriched horizon plays a critical role when considering the best possible prospective narrative.

Considering its role in Digital Poetics, exploration through creative narrative imagination could propose, for example, how semi-autonomous robots may impact jobs, household routines, child and elder care, laws, political systems, warfare, hazardous occupations, and so on. These imaginative variations mediated by narratives are important contributions to a denser deliberative narrative around why and how to deploy these types of technologies. This exploration also enables deliberators to consider a denser prospective narrative that accounts for more potential consequences of a given ethical decision related to a robot toy. Within Digital Poetics, inquiry into technology artifacts' features and applications is uniquely valuable as an instrument of deliberation prior to design and deployment, for technologies can radically affect people and society. Narrative imagination becomes an essential part of the development process when evaluating digital technologies that are not deployed yet but are in the process of design and implementation.

When considering phronetic deliberation through a lens of narrativity, Arendt's representative thinking is akin to a denser deliberative narrative. By thinking and feeling through fictional and historical narratives, decision-makers are able to augment the threads and voices represented in the deliberative narrative. This denser narrative is a fundamental step toward truly respecting others in their diverse needs and values since the final phronetic narrative requires the participation of diverse voices, not a sole deliberator. Whereas a denser narrative enhances the deliberative process through an enlarged exploration of the possibilities of meaning, Ricoeur's concept of critical practical wisdom catalyzes these narrative threads by placing them in a dialectical relationship that informs ethical deliberation. This characteristic of practical wisdom can foment a critique of ideologies and utopias created around new technologies. Are some technologies the result of valid utopian projects, or are they just manipulative ideologies guided by commercial interests? In chapter 6, we shall explore these far-reaching implications of phronetic deliberation in more detail when we analyze the institutional sphere of Digital Poetics.

Ethical Laboratories

Refiguration's critical role in the phronetic narrative leads us to reflect on the importance of fictional and historical narratives involving objects of deliberation. I turn our attention to the configuration moment of the triple *mimesis*, as narratives offer an important reservoir of productive imaginative variations that can contribute to the formation of good deliberators following the process of mimetic refiguration that expands semantic horizons and enriches cognitive models about the positive and negative dimensions of new technology development.

The need for ethical reflection that accounts for specific aspects of technologies from design to distribution and the human and ecological contexts that technology will affect has become more evident in recent years. David Kaplan argues, "We should examine how things are made, how they are used, and how they function in relation to broader cultural practices."[53] He contends that "philosophers of technology [should] examine the various ways that our technologies (plural) form the background, context, and medium for our lives, shaping our culture and the environment, altering patterns of human activity, and influencing who we are and how we live."[54]

On the basis of these suggestions, I highlight three main ideas that constitute an important context for our discussion. First, the use of the plural form of "technologies" stresses the attention to particular aspects of different artifacts. Although high-level discussion of "technology" as a whole, as proposed in the early stages of the development of the philosophy of technology by thinkers such as Heidegger and Marcuse, continues to be necessary, this discussion has to be complemented by a deeper understanding of the specificities of each technological innovation. Although my digital home assistant and coffee maker both fall under the umbrella of "technology" and share a spot on my kitchen counter, the ethical considerations surrounding them are markedly distinct. The coffee maker is a standalone device with a predefined interface, while the digital home assistant links directly to an AI engine driven by machine learning algorithms, hosted on servers potentially hundreds or thousands of miles away, presenting a unique set of ethical challenges. Second, we need to contextualize these technologies within cultural and social phenomena, current and future. And third, we must recognize the circle of interpretation between the creation and use of new technologies and the modification of the way we live and then create further technologies. This third point's emphasis on the way technology may affect the condition of human beings remains an underlying horizon of my argument.

As we discussed in chapter 2, John Wall advocated for an approximation between practical wisdom and *poiesis* that brings to the surface the poetic—the creative—potential of ethical decisions.[55] I agree with Wall's central argument, but I believe it can be extended not only to emphasize the poetic dimension of ethics but also the ethical dimension of poetics, which I apply here to technological innovation. The development of a social networking recommendation algorithm, for example, needs to address the possible ethical implications of using such algorithms. It is vital that creators of new technologies develop an ethical sensibility along with their technical and analytical skills, as I argued extensively when discussing the intrinsic relation between *praxis* and *poiesis* in the practice of software engineering.

To this end, any deliberative approach that removes technological artifacts from their contexts of use is insufficient. An ethical reflection on technologies should place them in the possible contexts in which they will be applied, and in this placement, imaginative variations appear as a fundamental ally to phronetic deliberation. Taking technologies as quasi-characters of historical and fictional narratives helps us explore and enhance the narrative understanding of the world and the possible meanings that certain computational and digital artifacts and architectures will bring to people and communities.

For example, it may become possible to create a technical means to broadcast live video streams directly to millions of social network users. Many of the so-called fifth-generation mobile networks (5G) that underlie this functionality will help patients who might benefit from remote and tele-mentored surgeries that leverage the high data transmission and low latency capacities of this new telecommunication infrastructures.[56] Nevertheless, serious consideration is needed during the development of such functionality regarding other potential malicious scenarios, such as when—in an actual historical case[57]—a serial killer broadcasts a terrible crime and leaves a digital trace of hundreds of copies. Digital plasticity allows the deliberative process to reshape the application of technology to ensure that the benefits of live streaming are accessible in a telemedicine scenario but that it is avoided as a means for disseminating videos of criminal acts.

My argument for the need and the potential of narratives, particularly in the digital context, to develop an ethical sensibility is reinforced by reflections on current philosophy. Several contemporary philosophers, including not only Ricoeur but Richard Kearney and Martha Nussbaum, have proposed and elaborated on the ethical implications of narratives. Kearney explores how stories shape the ethical understanding of different cultural traditions.[58] Stories model ethical understanding at various levels. From individual attention to the otherness of strangers to the call for mutual recognition to form large-scale communal entities, Kearney argues that narrative imagination has the potential to decentralize the ego and expand the horizon of meaning to account for other perspectives and experiences. Particularly given technologies' globalizing characteristics (digital ubiquity) discussed in chapter 1, Digital Poetics requires a deliberative process in which deliberators consider multicultural forms of use and diverse perceptions. "What narrative permits is the structuring of imagination in a way which propels it beyond its egotistical circle to a relation of analogy, empathy, or apperception (*Paarung*) with others."[59]

Nussbaum highlights the singularizing potential of narratives in ethics.[60] She contends that narratives propose an expansion of our understanding beyond a formal and deontological approach to include the "exemplary" persuasiveness of literary and oral stories. This singularizing power is especially necessary to take into account the perspective of minority groups or types of technology users who are not usually considered in the main use cases of digital artifacts. Extending her argument, in her book *The New Religious Intolerance* Nussbaum outlines three fundamental criteria for the development of an ethical and tolerant attitude.[61] In addition to political principles that, first, respect all citizens and, second, require rigorous critical thinking, she highlights third the fundamental importance of cultivating inner eyes, "the imaginative capacity that makes it possible for us to see how the world looks from the point of view of a person different in religion or ethnicity."[62] Not surprisingly, later in her book Nussbaum brings this participatory imagination closer to literary narratives that foster an inner eye by providing different perspectives and helping us to imagine experiences from diverse perspectives. Citing Daniel Batson's studies, Nussbaum explains that literary narratives have the effect of developing an attentive and caring behavior toward others, developing an "empathic imagination."[63] Adelaide Fins highlights that Nussbaum's attention to narrative imagination is akin to Ricoeur's concept of solicitude in the focus on recognition of fragile and vulnerable narratives that need to be taken into account when making ethical decisions: "[Nussbaum] invites us to think about the complexity of reality, not with reference to abstractions, but through attention to ordinary and vulnerable life."[64]

The singularizing potential stressed by Nussbaum is tightly related to practical wisdom and phronetic deliberation on digital matters. Narratives have the potential to connect new technology developers and institutional deliberators not only to abstract moral norms but to the lives of future users of those technologies through the imaginative role of fictional narratives, including those whose representativeness as participants in a technological society may be more marginalized and vulnerable.

As for Ricoeur, we noted earlier his apparent resistance to the development of narrative ethics. Yet in *Time and Narrative*, we also find him questioning the hypothetical ethical neutrality of narratives, which he claims attempts to suppress "one of the oldest functions of art, that it is an ethical laboratory where the artist pursues through the mode of fiction experimentation with values."[65] Later in the same volume, Ricoeur points out that narratives are "that extraordinary laboratory of the probable constituted by the paradigms of emplotment."[66] These imagined paradigms of emplotment are always linked to possible ethical implications that vary according to plot schemata. In *Oneself as Another*, Ricoeur extends the idea of the narrative laboratory, stating: "Literature is a vast laboratory in which we experiment with estimates, evaluations, and judgments of approval and condemnation through which narrativity serves as a propaedeutic to ethics."[67] In what sense are narratives a propaedeutic for ethical deliberation? Asynchronous plurality captures exactly this feature of narratives. Deliberators broaden their horizons of meaning through the mimetic process and apply these refigurations in phronetic deliberation. Narratives can sharpen sensitivity to ethical issues that an exclusively technical worldview ignores.

Narrative Ethics and Digital Technology: The Example of Science Fiction

If we link the function of narratives to the ethical implications of digital technologies, we are led quite naturally to the genre of science fiction. Science fiction performs precisely a kind of creative, imaginative variation by placing technologies in different contexts and exploring possible personal, interpersonal, and social implications that certain technological advances may create. Eileen Botting, for instance, explores how Mary Shelley's Frankenstein is the "prototype for all speculative and science fiction on the ethics of the artificial creation, modification, and transformation of human life."[68] Within narrative deliberation, undoubtedly science fiction narratives offer a particular affinity with Digital Poetics because, in many cases, they deal with ethical issues directly related to technological advances.[69] This is the case, for example, with science fiction narratives that offer perspectives on the possible implications of developments in artificial intelligence. "Narratives of intelligent machines matter because they form the backdrop against which AI systems are being developed, and against which these developments are interpreted and assessed."[70]

When assessing the association between fictional narratives and science fiction in the context of new technologies, some critical considerations need to be offered. First, the definition of the science fiction genre is complex and involves a conceptual controversy that goes beyond the context of the present work. For our purposes, regardless of the response within literary studies to the issue of genre classification, any work that discusses how a real or possible technological artifact affects or may affect our being-in-the-world horizon brings considerations that are potentially relevant to an ethical perspective. For example, in *Klara and the Sun*,[71] Kazuo Ishiguro addresses issues of artificial companionship, artificial consciousness, and digital immortality related to possible developments in the fields of robotics and artificial intelligence. Although Ishiguro's novel may not be classified as science fiction, it is an excellent example of the potential of narratives involving digital technologies to arouse ethical sensitivity.

The fluidity between the "possible" and the real through fictional narratives is valuable to our discussion even if a given work does not make it to the science fiction shelf due to certain taxonomic criteria. Even though works in science fiction seem like a good starting point for this discussion, one needs to remain open to any fictional narrative that explores the "possible" of technological innovations. In his article on the potential of prospective narratives to disrupt and propose new possibilities for political identities, George Taylor explores the dialectics between the real and the possible mediated through Ricoeur's development of the concept of figuration: "[figuration] intertwines the 'real' and the 'unreal.' The 'real' is not a static category, and fiction can offer us insight into the 'real.'"[72] This dynamicity of the "real" investigated through the lens of the fictional "possible" is attested to every day by the ways in which technologies operate the conversion of the "possible" into the "real."

Another prudent constraint on focusing too much on science fiction in narrative deliberation within a Ricoeurian framework emerges from Ricoeur's distancing himself from a particular use of science fiction scenarios when discussing the problems associated with narrative identity in relation to Derek Parfit's puzzling cases.[73] This issue seems closely linked to the context of application to the question of personal identity and not to the science fiction genre itself. In his debate with Parfit on personal identity, Ricoeur wonders whether the transformation of the human bodily condition to act and suffer into a mere contingent variable does not endanger the possibility of ethical imputability. Violation of this existential condition carries with it the risk that both ethical principles themselves and moral laws will become empty, given the impossibility of imputation to a flesh-and-blood person,[74] which is essential to our argument. Parfit's puzzling cases reduce personal identity to sameness through the separation of body and mind in a manner that does not consider that the body is not just any body, but my own body (*corps propre*). The non-violation of fundamental human existential characteristics should be an important criterion used to evaluate the possible contributions of fictional narratives to an ethical understanding that remains practically relevant and not merely theoretical and speculative.

To finalize these possible objections to the use of science fiction, I want to address an argument that goes beyond one specific genre and relates to all types of narratives applied to ethical deliberation. In a volume edited by Morny Joy on narrative theory,[75] Ricoeur has made it clear that imagination and memory in the form of metaphors and narratives can complement an ethical proposition, but they cannot replace the whole panoply of an ethics that, as we have discussed, involves the complete arc of his "Little Ethics" from the desire for the good life, through the test of deontological norms, to practical wisdom as critical phronesis. Ricoeur's concern does not apply to our development of phronetic deliberation, since we have, from the very beginning, taken this comprehensive ethical perspective into consideration.

While I agree with Ricoeur's position that a narrative approach alone is insufficient for fundamental ethics, I continue to maintain, as I have argued, that paradoxically, he has overlooked the potential of his narrative theory for applying critical practical wisdom to conflictual situations, with which he culminates his "Little Ethics." As I have suggested, the deliberative process must incorporate ethical principles as testing constraints to prospective narratives. This obligation should lead us to think about potential paraphrases of the Kantian categorical imperatives from a narrative perspective: choose only the prospective narratives in which everyone can be protagonists in a life fulfilled with and for others in just institutions. Or: choose only prospective narratives in which everyone is treated as ends and never means in the plot narratives.

In concluding this section, we may summarize its trajectory as follows. Beginning with the idea of imaginative variation derived from Husserl's phenomenological description, Ricoeur expands the reach of imaginative variation and emphasizes its innovative potential through productive imagination. This productive imaginative process offers a redescription of the world and opens windows to explore other ways in which technologies create diverse forms of being in the world: "Through fiction and through poetry new ways of being-in-the-world are opened up in the midst of reality."[76] Thus, fiction and poetry—and, I have argued, narratives in general—are expressions of creative imaginative variations that are particularly integral to phronetical deliberation, because they apply innovative potential to the emplotment of actions. Ricoeurian narrative theory emphasizes this relationship between the poetic configuration of narratives with the domain of action and ethics through a threefold *mimesis* iterative cycle. Narratives become springs for creative narrative variations on the ethical implications of the objects of deliberation.

Semantic Innovation

While I have already discussed the role of creative imaginative variation in bringing new and different contexts to bear on phronetic deliberation, I want to turn now more explicitly and intensively to the role of imagination in the process of phronetic judgment itself. The narrative approach enriches the deliberative process in another crucial way, as it provides a form of creative imagination that explores new possibilities in the face of uncertainties in new technology design. Practical wisdom consists of a creative activity that invents new possibilities of action.[77]

Ricoeur pays special attention to the limitations of exclusively logical-rational processes taken as the sole source of solution to ethical questions. Although these processes are a necessary condition, a mandatory step in the development of ethical problems, they run into the "tragic action"[78] that requires new and imaginative solutions to seemingly insoluble ethical dilemmas taken strictly from a logical deductive framework. Tragic action demands a poetic answer that transcends the aporia of the logical impasse through the imaginative dimension of practical wisdom. The prospective narrative is the result of a novelty brought to light by the deliberators' creative imaginative variations. The passage from the *a-poria* (literally, a situation where there is no way out) that precipitated the impasse of ethical action to the *eu-poria* (literally, a good way out) of good deliberation is the result of the imagination of new action schemata that address the particularity of the situation, the respect for others, and the due regard for the ethical principles taken as its basis.

In his book *Les puissances de l'imagination*, Jean-Philippe Pierron argues for something very close to this connection between imagination and practical wisdom. He investigates how imagination assists in the configuration of practical reason: "To this end, we will focus on the innovation brought about by the creativity of the ethical imagination in human action."[79] Pierron offers three main reasons for supporting the importance of a poetics of action that consistently integrates imagination with moral proposals. First, an ethical imagination offers a rejoinder to an instrumental rationality that "bureaucratizes" existence and an ethical life that now dominates our normative context as a whole. The modern context reduces ethics to a technique and practical resolution merely to a logical solution. Second, there is today an excessive valuation of the emotional aspect of judgment: "a culture of emotion exacerbating sensational enthusiasm for the pain of the world seems to dispense easily with the necessary practical mediation that responds to the suffering or joy of men."[80] Finally, Pierron diagnoses the exhaustion of a line of research that seeks exclusively in rational processes the core of our reasons for acting. He asks rhetorically, "Is there any possibility of living in a situation without a capacity for practical innovation, inventiveness and creative imagination?"[81]

This positive perspective on imagination has not prevailed throughout the history of philosophy. On the contrary, for example, before the shift in perspective offered by thinkers such as Arendt and Ricoeur, Pascal and Malebranche deplored the use of imagination in the ethical domain. It was, for them, the source of error and the antonym of the straight rational thought that forms "good" ethics. Kant also rejected the use of imagination to find moral principles in the context of moral schematism. For Kant, imagination does not have the same importance in the practical domain as it does in the theoretical realm and in knowledge production. Imagination retains merely a presentational role in contexts of moral law as a "moral propaedeutic": "The imagination can therefore have its place when it comes to thinking about morality, not of reasonable beings, but in man, and stimulating in him an emotional dynamic that would have the educational role of a moral propaedeutic."[82]

By contrast, with Arendt and Ricoeur, imagination regains full status as a citizen in the realm of ethics and morals. They broke with the tradition of ethical imagination by effectively moving from the theory of knowledge to action. In Arendt, the main scope is political; in Ricoeur, ethics unfold in institutions. Arendt argues that imagination is linked to an anthropology of the "imaginer," both as agents and sufferers. Imagination is one of the most remarkable human capacities. She sees in the imagination the powerful ability to generate new beginnings on the plane of action. Imagination becomes one of the main features of action, in contrast to the concepts of work and labor. In this sense, the unpredictability of action demands imagination to constantly find new ways of acting: "the new therefore always appears as a miracle. . . . To act is to begin, to act is to innovate."[83] Arendt's central argument seeks to stop envisaging imagination in opposition to other forces such as reason, will, and affections. Imagination must have a transversal role in moral judgment that impacts both internal and external aspects of the moral agent and, in our context, phronetical deliberation. Internally, imagination increases the agent's inventive capacity. It also allows for a greater capacity for deliberators to distance themselves from the restrictions of a solely logical and empirical description of "objective reality" and to foster an attitude of critical thinking about constituted reality. Externally, imagination manifests itself when developing the anticipation that motivates a new project's inception. Imagination works as a motor that initiates the action process.[84] In our more specific context, imagination helps inspire the desire to create or change technologies so that they promote the common good.

Pierron's portrayal of the role of imagination in Ricoeur's ethical proposal obviously interests us in a special way. Pierron emphasizes that for Ricoeur, ethical action appears as a work of the imagination on three levels.[85] The first level is the plane of choice. At this level, imagination primarily has a projective function. It is a necessary part of the project's elaboration, the means, and the ends that give unity to the project. On a second level, imagination plays the role of explaining and nurturing our motivations, and on this level, it is more connected to the inner life of the self. Through imagination, "we see ourselves as if," "we represent ourselves as if," "we become characters as if." At stake here is a fundamental aspect of the capacity for refiguration of the self (and in our particular case of the *phronimos*). The mimetic refiguration depends on this imaginative capacity for the "as if" as though the self were the real or fictional other. The third level is closer to action. It deals with questions such as "what can we do given that," "what will I tell you, what can I do." Imagination thus makes palpable our "power of," and it unites the person of dreams to the person of action. This level is fundamental in the phronetic deliberation process as it relates to the continuous exercise of thinking about possible alternatives to the aporias of ethical dilemmas. These imaginative ethical variations will need to pass through the sieve of dialogue, the contestations of debate, and the filtering of ethical principles, norms, and guidelines. Still, imaginative ethical variations are fundamental for overcoming ethical dilemmas, as they are the primary sources of new possible paths that promote ways of advancing prospective narratives.

Following this productive intellectual journey with Pierron, through which we reflected on the relationship between ethics and imagination, we conclude the arc of this section by returning to the relevance of the creative imagination in the context of narrative deliberation. We will do so under the auspices of Ricoeur's concept of semantic innovation. Ricoeur brings his work on metaphors and narratives together around the common core of the creative imagination that can carry out innovative syntheses from apparently diverse elements. Within the Kantian epistemological context, it is about applying a schema to account for a heterogeneous set of elements: a link between logical and sensible components or between the verbal and the nonverbal.[86] Specifically regarding narratives, Ricoeur says that semantic innovation occurs "in the invention of another work of synthesis—a plot. By means of the plot, goals, causes, and chance are brought together within the temporal unity of a whole and complete action."[87] It is precisely this innovative synthesis that springs from creative imagination and gains materiality in the form of a plot that lies at the heart of our hypothesis about the potential of narrative theory for the composition of prospective deliberative narrative.

The creative capacity of the narrative imagination supports the task of practical wisdom to turn aporetic situations into good deliberations. Using the narrative framework, we can see phronetic deliberation as a process of re-schematization. The retrospective narrative is a schema of the state of affairs involving the deliberative object that does not (or still does not) promote a good life in common. The prospective narrative is the exercise of re-schematization that explores new possible configurations of actions, characters, and events to produce a new narrative scheme in the future that establishes or re-establishes the possibility of the good life. The genius involved in practical wisdom is, thus, epistemologically related to the poetic capacity to create new narrative schemas through new emplotments of actions.

Laws, norms, and guidelines are more abstract schemas that function as templates for ethical deliberations. They propose meta-narratives or practical constraints for prospective narratives that need to be instantiated to account for the contextual peculiarities and the respect for people involved in the ethical dilemma in question. The narrative re-schematization of practical wisdom is challenging because it must shape the phronetic narrative on the basis of horizontal (syntagmatic) and vertical (paradigmatic) constraints. Horizontally, the emplotment of the prospective narrative must follow the narrative logic of the retrospective narrative. The scheme of the state of affairs that the retrospective narrative configures imposes limitations on what can be told through the prospective narrative to be narratively coherent. Vertically, universal moral principles and the respect owed to every person involved in deliberation also limit the re-schematization of narrative deliberation. To make things even more complex, as Ricoeur highlights in *Oneself as Another*, often these principles are contradictory when applied to a particular ethical problem, again characterizing for Ricoeur the tragic nature of action.

Faced with this multidimensional puzzle, practical wisdom resembles narrative intelligence's semantic innovation by proposing a creative schema of things and actions that increases the possibility of a good life in common and attends—to the greatest possible extent—to all syntagmatic and paradigmatic constraints.

Poetic Universals

Aristotle famously affirmed in his *Poetics* that "poetry is more philosophical and more elevated than history, since poetry relates more to the universal, while history relates particulars."[88] Ricoeur comments on this passage, highlighting the potential universal applicability of fictional narratives: "Aristotle did not hesitate to say that every well-told story teaches something; even more, he said that stories reveal universal aspects of the human condition and that, therefore, poetry is more philosophical than the history of historians, who are too dependent on anecdotic aspects of life."[89] Fictional narratives' potential to reveal *poetic universals* of action that can inform agency and ethical decisions intrigued Aristotle and Ricoeur. In this section, we will pursue this intuition regarding fictional narratives and extend it to investigate the application of this dialectic between narrated plot and poetic universals in the context of prospective narratives. First, we will examine how fictional narratives are indeed ethical sources for deliberations about digital technologies on the basis of the poetic universals that emerge from these narratives. Then we will see how poetic universals can help us better grasp how a narrative approach emphasizes the intense dialectical process between practical wisdom and general moral principles.

Poetic Universals and Fictional Narratives

As suggested by Aristotle and then extensively explored by Ricoeur in *Time and Narrative,* poetic universals are rather different from the universals of logical and theoretical thought. The kind of universality implied by narratives is that of a universality tied to the sequence of actions that, as Ricoeur points out, is also linked to practical wisdom:

> The kind of universality that a plot calls for derives from its ordering, which brings about its completeness and its wholeness. The universals that a plot engenders are not Platonic ideas. They are universals related to practical wisdom, hence, to ethics and politics. A plot engenders such universals when the structure of its action rests on the connections internal to the action and not on external accidents.[90]

Given the innumerable interpretive possibilities this citation evokes, I would like to highlight just one aspect that seems fundamental for building the intended bridge between narrativity and phronetic deliberation and their applicability to the underlying contexts of technologies. Narratives, through plots, models, or schemas, apply creative imaginative variations to the possible ways in which ethical principles can turn into particular sequences of actions. However, these action schemata maintain the trait of universality that allows us to speak of poetic universals. They develop a level of abstraction that lies between a general principle applicable to any action and levels of poetic realization of such principles within a particular contextual framework of action. Such in-betweenness in one direction makes it easier to abstract the plot even further and formalize it as a general principle, and in the other direction, it makes poetic universals closer to potential future particular situations of aporias that require phronetic deliberation.

Richard Kearney comments on the relationship between narrative understanding and the practical wisdom of moral judgments, noting that narratives relate aspects of human action to the ideal of a fulfilled or happy life.[91] He recalls that when the Greeks, for example, explained courageous actions, they told stories about Achilles; when they spoke about daring, they told about the feats of Prometheus. Finally, he talks about how these lessons of narrative constitute the poetic universals of which we speak.

In this sense, fictional technology narratives create imaginative variations that offer schemas of action configured at a level between ethical universals and the contexts in which certain technological artifacts will affect our way of being in the world. By expanding the spectrum of ethical sensibilities, such poetic universals enable the audience to engage in deeper and denser phronetic deliberation because they point to possible effects that the creation or use of a given technology will have. Such deliberation also allows us to make proactive changes to the technological artifact itself or to the infrastructure that supports it to minimize possible negative effects.

Thus, the development of a particular wireless communication technology that enables instant communication between one person and millions of others should take into account possible narratives in which this feature is used to publish real scenes of violence that could have negative effects on the lives of victims, and also viewers, quite possibly nurturing ideologies of hatred and terror. Fictional narratives occupy this space of possible meanings between the universal notion of not disrespecting another's life and the use of a specific communication technology. Such imaginative variations can become part of the prefigured understanding of deliberators who will have to decide the use and development of a particular technology that involves wireless communication. The poetic universals emerging from fictional narratives enhance the understanding of meanings, consequences, and perspectives of the passage from universal principles to the particulars of a given ethical decision.

This formative role of poetic universals explains Ricoeur's suggestion that fictional narratives are a propaedeutic for ethical deliberation in the context of critical phronesis.[92] The dialectic between universals and particulars, mediated by the poetic universals proposed by narratives, is a fundamental characteristic of practical wisdom, which operates precisely in the area of attention to the singular characteristics of a specific narrative upon which it is based. This characteristic of phronetic thought is key to properly exploring the implications of new technologies in their application contexts and central as well to moving the focus from a general category to the multiple and fast-changing forms of different technological artifacts, of technologies in the plural. In the following subsection, we will advance this argument to consider poetic universals emerging not only from fictional narratives but also from prospective narratives. These reflections will invite us to consider the more profound and nuanced ways in which deliberative narratives interact.

Poetic Universals and Prospective Narratives

As we have emphasized, one of the fundamental characteristics of practical wisdom is that it is not just a straightforward, blind application of general moral principles or norms. Something new always exists in every phronetic deliberation that responds to the specifics of the ethical situation in question and the respect due to the people involved. The degree of ethical innovation that phronetic deliberation demands will vary depending on how much existing rules and laws are adapted to the ethical dilemma at hand. Yet one common attribute of digital technologies is that by creating new experiences and relationships in the world, they may also trigger the need for unprecedented deliberation. For instance, twenty years ago, parents did not need to deliberate about the number of hours their children should be allowed to use mobile devices at home. Given the digital ubiquity of these devices currently, however, parents are now faced with a new ethical dilemma that requires an imaginative solution and proposes a prospective narrative such as: "my child can use his or cell phone for an hour after doing homework and never while we're at the dinner table."

Each prospective narrative advances an imaginative solution to a moral dilemma that often does not precisely coincide with general rules. However, this process is not merely unidirectional in the sense that general rules and principles exist statically at a single level of abstraction from which phronetic deliberations emerge, act in that particular instance, and leave no trace. Instead, ethical deliberations, norms, and principles are dynamic, cyclically related, and organized into multiple levels of cognitive categories.

Thus, new phronetic deliberations configure inchoate poetic universals that affect rules and norms, altering them, creating new ones, or proposing intermediate levels such as guidelines or standards of conduct. What we have in mind is akin to precedents within legal procedures as they reshape, complement, or create new deliberative standards that will impact new instances of deliberation. In a dialectical relationship with general principles, norms, and laws, practical wisdom enlivens deliberative processes and the legal structures themselves. For example, within Digital Poetics developers' sphere of deliberation, technology designers can propose a prospective deliberation in which it is possible for parents to set a maximum limit of hours they deem healthy for their children to interact with the device. This forward-looking narrative for the case of a specific operating system can emerge as a poetic universal. Parts of the deliberations derived from particular situations raise their levels of abstraction and become paradigms for future deliberations in the form of best practices, guidelines, professional standards, and can, eventually, become part of laws and inform revised general principles. Poetic universals, born out of a particular phronetic deliberation, grow in their level of abstraction and guide other deliberations in transforming the ways in which a universal principle is realized in intermediate ethical and legal categories.

A narrative approach to the deliberative process captures this dynamic nature of practical wisdom through the concept of poetic universals that emerge from particular narratives. As we have discussed, this movement, which is essential in fictional narratives, is also operative in prospective narratives, since each deliberation over a particular ethical dilemma might contain poetic universals applicable to a range of similar situations that relate to this new context that digital technologies bring up.

Figure 3.1 attempts to synoptically capture part of the conceptual relationships in this cycle between poetic universals and phronetic deliberations. The first moment, (1) in figure 3.1, highlights the poetical universals that emerge from the plots of fictional narratives. For instance, a story may offer a poetic universal message of helping those in need, even where there is no previous relationship of friendship, belief, or kinship. In the second moment (2), such a poetic universal might serve as input for the configuration of an unprecedented prospective narrative born from an ethical dilemma entirely different from the particular circumstances narrated, for example, how a specific autonomous vehicle being designed should behave when interacting with an accident involving a human. The third moment (3) highlights that any prospective narrative can also generate new poetic universals that will impact future phronetic deliberations, thus closing the productive cycle proposed here.

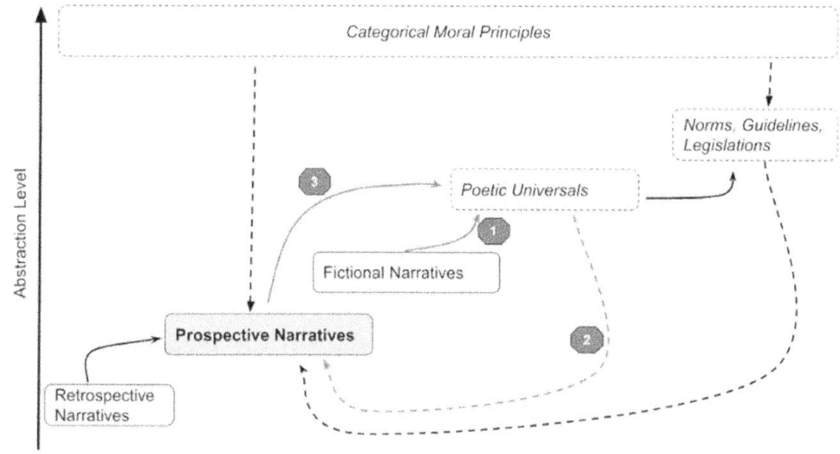

Figure 3.1 A Synoptic View of the Narrative Deliberation Conceptual Framework.

With this, we conclude the reframing of the phronetic deliberation process based on Ricoeur's narrative theory. We organized the conceptual edifice of this chapter based on the narrative deliberation framework that consistently articulates a diversified analysis of the past and the public debate that emerges from the retrospective narrative with innovative solutions to ethical dilemmas through narrative imagination guided by moral principles. The narrative approach also gives us a key insight through the concept of narrative density about how diverse and nuanced a deliberation is through the concept of narrative density. The synchronous and asynchronous phronetic plurality that arises from the narrative configuration and refiguration highlights the importance of triple *mimesis* for a new understanding of phronetic deliberation. The narrative understanding of the nuances of the semantic web of actions proves to be a crucial ally in the essential movement of practical wisdom to contextualize general principles in particular situations of human action and interaction. Finally, the temporal schematism typical of narrative imagination becomes an empowering capability for practical wisdom to create a *euporia* out of the paralyzing *aporia* of the simple application of moral principles to complex, sometimes tragic, particular situations.

In the following three chapters, we will apply narrative deliberation to the various contexts in which ethical decisions about the design, development, application, and regulation of digital technologies need to be made. We will organize these contexts into three spheres of deliberation representing forms of agency concerning narrative deliberation: the sphere of developers, users, and institutions. Two premises guide this organization into spheres of deliberation. First, a comprehensive ethical proposal to address the implications of digital technologies cannot be reduced to developers' decisions. The sphere of developers is fundamental but not exclusive. The ubiquity of digital technologies demands an equally ubiquitous deliberative effort. Second, there are several mutual implications between the three spheres of deliberation. Each sphere offers an angle from which to approach Digital Poetics, but when we explore the issues of one sphere, we always find countless dependencies and repercussions in the other spheres.

Notes

1 Paul Ricoeur, "À La Gloire de la Phronesis (Éthique a Nicomaque, Livre VI)," in *La vérité pratique: Aristote, Ethique à Nicomaque, Livre VI*, edited by Jean-Yves Chateau (Paris: Vrin, 1997), 13–22.

2 Elsewhere in his work, Ricoeur makes it clear how this anthropological option, which is closely related to the power to act and the effort to exist, and his readings of Jean Nabert and Baruc Spinoza are fundamental to the whole structure of Ricoeurian practical philosophy. See, for instance, Paul Ricoeur, *Réflexion Faite: Autobiographie Intellectuelle* (Paris: Édition Esprit, 1995), 106.

3 Ricoeur, "À La Gloire de la Phronesis," 17. My translation.

4 Paul Ricoeur, "La Raison Pratique," in *Rationality To-Day*, edited by T. F. Geraets, 225–48 (Ottawa: University of Ottawa Press, 1979), 235. My translation.

5 Ricoeur, "À La Gloire de la Phronesis," 20. My translation.

6 Paul Ricoeur, *Oneself as Another* (Chicago: University of Chicago Press, 1994), 290–91.

7 Ricoeur, *Oneself as Another*, 269.

8 Paul Ricoeur, *Time and Narrative, Volume 1* (Chicago: University of Chicago Press, 1984), 66.

9 Ricoeur, *Oneself as Another*, 172.

10 Ricoeur, *Time and Narrative, Volume 1*, 56.

11 See, for instance, Ricoeur, *Oneself as Another*, 246; Paul Ricoeur, *On Psychoanalysis* (New York: Wiley, 2012), 190; Paul Ricoeur, "The Text as Dynamic Identity," *Identity of the Literary Text*, n.d, 177; Paul Ricoeur, *Figuring the Sacred: Religion, Narrative, and Imagination* (Minneapolis: Fortress Press, 1995), 239.

12 Paul Ricoeur, "Life in Quest of Narrative," in *On Paul Ricoeur. Narrative and Interpretation*, edited by D. Wood (London and New York: Routledge, 1991), 20–33.

13 Ricoeur, "Life in Quest of Narrative," 23.

14 Ricoeur, *Time and Narrative, Volume 1*, 55.

15 Ricoeur, *Time and Narrative, Volume 1*, 56.

16 Paul Ricoeur, "Narrative Time," *Critical Inquiry* 7, no. 1 (October 1, 1980): 169–90, 174.

17 Peter Kemp discussed Ricoeur's take on a narrative approach to ethics and acknowledged that "[in Ricoeur's] published expositions directly concerned with ethics, Ricoeur is hesitant to take a step toward a narrative understanding of ethics." Peter Kemp, "Toward a Narrative on Ethics: A Bridge between Ethics and the Narrative Reflection of Ricoeur," Translated by Craig Dilworth. *Philosophy & Social Criticism* 14, no. 2 (April 1, 1988): 179–201.

18 Ricoeur, *Oneself as Another*, 164–65.

19 Richard Kearney, "Parsing Narrative—Story, History, Life," *Human Studies* 29, no. 4 (2006): 477–90, 482.

20 Kearney, "Parsing Narrative—Story, History, Life," 486.

21 Paul Ricoeur, *Time and Narrative, Volume 3* (Chicago: University of Chicago Press, 1988), 184.

22 Richard Kearney, "Narrative Imagination: Between Ethics and Poetics," *Philosophy & Social Criticism* 21, no. 5–6 (1995): 173–90, 182.

23 Ricoeur, *Time and Narrative, Volume 3*, 143.

24 Porter Abbott, *The Cambridge Introduction to Narrative* (Cambridge: Cambridge University Press, 2018), 52.

25 Paul Ricoeur, *Hermeneutics and the Human Sciences: Essays on Language, Action and Interpretation* (Cambridge: Cambridge University Press, 1981), 215.

26 The focus here is on the development or enhancement of new and existing technologies. There are certainly other cases in which specific actions and individuals should be judged based on their creation or use of digital artifacts.

27 Ricoeur, *Time and Narrative, Volume 1*, 75.

28 Paul Ricoeur, *Memory, History, Forgetting* (Chicago: University of Chicago Press, 2004), 380.

29 Ricoeur, *Oneself as Another*, 164–65.

30 Ricoeur, *Oneself as Another*, 178.

31 Although I cannot take the time in this volume to provide more detailed clarification, I would argue that the relationship between retrospective and prospective narratives can also fruitfully be compared to Geertz's concepts of "model of" and "model for." If narratives are temporal and linguistic models of actions and events, retrospective narratives are "models of," and therefore fundamentally try to describe the actions and events that led to the moment of deliberation. On the other hand, prospective narratives tend to be a "model for" that explore and investigate possibilities in which actions and events will unfold in the future. Clifford Geertz, *The Interpretation Of Cultures* (New York: Basic Books, 1973), 93.

32 Ricoeur, *Oneself as Another*, 115.

33 Plato, *The Statesman* (Cambridge: Cambridge University Press, 1995), 284a–e.

34 Ricoeur, *Oneself as Another*, 141.

35 Ricoeur, *Time and Narrative, Volume 1*, 52.

36 Ricoeur, *Oneself as Another*, 115.

37 Jitendra Nath Mohanty, "Method of Imaginative Variation in Phenomenology," *Thought Experiments in Science and Philosophy* (1991): 261–72.

38 Mohanty, "Method of Imaginative Variation in Phenomenology," 265.

39 Paul Ricoeur, *Husserl: An Analysis of His Phenomenology* (Evanston: Northwestern University Press, 1967), 42.

40 Ricoeur, *Husserl: An Analysis of His Phenomenology*, 42.

41 Edmund Husserl, *Cartesian Meditations: An Introduction to Phenomenology*. Translated by Dorian Cairns (Netherlands: M. Nijhoff, 2012), 262.

42 Clark Moustakas, *Phenomenological Research Methods* (Thousand Oaks: SAGE Publications, 1994), 60.

43 Aristotle, *Poetics*. Translated by Stephen Halliwell. Volume XXIII. Loeb Classical Library (Cambridge, MA: Harvard University Press, 1995).

44 Paul Ricoeur, *Fallible Man* (New York: Fordham University Press, 1965), 112.

45 Paul Ricoeur, "Intellectual Autobiography," in *The Philosophy of Paul Ricoeur*, edited by Lewis Edwin Hahn (Chicago: Open Court, 1995), 44.

46 Eric Chown and Fernando Nascimento, "Software and Metaphors," in *Interpreting Technology: Ricoeur on Questions Concerning Ethics and Philosophy of Technology* (Lanham: Rowman & Littlefield, 2021), 229.
47 Donald A. Schön, "Generative Metaphor: A Perspective on Problem-Setting in Social Policy," *Metaphor and Thought* 2 (1979): 137–63.
48 Paul Ricoeur, *Lectures on Ideology and Utopia*. Edited by George Taylor (Ithaca: Columbia University Press, 1986), 28.
49 Ricoeur, *Lectures on Ideology and Utopia*, 299.
50 Ricoeur, *Lectures on Ideology and Utopia*, 270.
51 Ricoeur, *Time and Narrative, Volume 1*, 68.
52 Hannah Arendt, *Between Past and Future: Eight Exercises in Political Thought* (New York: Penguin Books, 1993), 200–201.
53 David Kaplan, *Readings in the Philosophy of Technology*, 2nd Edition, Volume 37 (Lanham: Rowman & Littlefield Publishers, 2014), Introduction, Kindle Edition.
54 Kaplan, *Readings in the Philosophy of Technology*, Introduction, Kindle Edition.
55 John Wall, "Phronesis, Poetics, and Moral Creativity," *Ethical Theory and Moral Practice: An International Forum* 6, no. 3 (September 1, 2003): 317–41.
56 Krunal Pandav, Austen G. Te, Nir Tomer, Sujit S. Nair, and Ashutosh K. Tewari, "Leveraging 5G Technology for Robotic Surgery and Cancer Care," *Cancer Reports* 5, no. 8 (August 2022): e1595.
57 Ryan Mac, Kellen Browning, and Sheera Frenkel. "The Enduring Afterlife of a Mass Shooting's Livestream Online," *The New York Times*, May 19, 2022, https://www.nytimes.com/2022/05/19/technology/mass-shootings-livestream-online.html.
58 Kearney, "Narrative Imagination: Between Ethics and Poetics," 173–90.
59 Kearney, "Narrative Imagination: Between Ethics and Poetics," 184.
60 Martha Nussbaum, *Poetic Justice: The Literary Imagination and Public Life* (Boston: Beacon Press, 1997).
61 Martha Nussbaum, *The New Religious Intolerance* (Cambridge, MA: Harvard University Press, 2012).
62 Martha Nussbaum, *The New Religious Intolerance* (Cambridge, MA: Harvard University Press, 2012), 3.
63 C. Daniel Batson, Shannon Early, and Giovanni Salvarani, "Perspective Taking: Imagining How Another Feels Versus Imaging How You Would Feel," *Personality & Social Psychology Bulletin* 23, no. 7 (July 1, 1997): 751–58.

64 Adelaide Fins, "Repenser L'éthique à Travers L'imagination Narrative et Littéraire Dans La Pensée de Paul Ricœur et de Martha Nussbaum," *Bulletin D'analyse Phénoménologique(Actes 10)* XII, no. 2 (2017): 480 (my translation).
65 *Ricoeur, Time and Narrative, Volume 1*, 59.
66 *Ricoeur, Time and Narrative, Volume 1*, 184.
67 Ricoeur, *Oneself as Another*, 115.
68 Eileen Hunt Botting, "Frankenstein and the Question of Children's Rights After Human Germline Genetic Modification," in *Reproductive Ethics II* edited by Lisa Campo-Engelstein and Paul Burcher (Cham: Springer International Publishing, 2018), 9–24.
69 Beleicia B. Bullock, Fernando L. Nascimento, and Stacy A. Doore, "Computing Ethics Narratives: Teaching Computing Ethics and the Impact of Predictive Algorithms," in *Proceedings of the 52nd ACM Technical Symposium on Computer Science Education* (New York: Association for Computing Machinery, 2021), 1020–26.
70 Stephen Cave, Kanta Dihal, and Sarah Dillon, *AI Narratives: A History of Imaginative Thinking about Intelligent Machines* (Oxford: Oxford University Press, 2020), 7.
71 Kazuo Ishiguro, *Klara and the Sun* (London: Faber & Faber, 2022).
72 George H. Taylor, "Prospective Political Identity," in *Paul Ricoeur in the Age of Hermeneutical Reason*, edited by Roger W. H. Savage, 123–37 (Lanham: Lexington Books, 2015), 124–25.
73 Ricoeur, *Oneself as Another*, 150–51.
74 Ricoeur, *Oneself as Another*, 150–51.
75 Morny Joy, ed., *Paul Ricoeur and Narrative: Context and Contestation* (Calgary: University of Calgary Press, 1997).
76 Paul Ricoeur, "Philosophical Hermeneutics and Theological Hermeneutics," *Studies in Religion. Sciences Religieuses* 5, no. 1 (1975): 13.
77 Ricoeur, *Oneself as Another*, 269.
78 Ricoeur, *Oneself as Another*, 241–49.
79 Jean-Philippe Pierron, *Les Puissances de L'imagination: Essai Sur La Fonction éthique de L'imagination* (Paris: Les éditions du Cerf, 2012), 12. My translation.
80 Pierron, *Les Puissances de L'imagination*, 14. My translation.
81 Pierron, *Les Puissances de L'imagination*, 18. My translation.
82 Pierron, *Les Puissances de L'imagination*, 106. My translation.
83 Pierron, *Les Puissances de L'imagination*, 117–18. My translation.

84 Pierron, *Les Puissances de L'imagination*, 118. My translation.

85 Pierron, *Les Puissances de L'imagination*, 126. My translation.

86 Paul Ricoeur, *A Ricoeur Reader: Reflection and Imagination* (Toronto: University of Toronto Press, 1991), 126.

87 Ricoeur, *Time and Narrative, Volume 1*, xi.

88 Aristotle, *Poetics*. Translated by Demetrius Longinus, 1451b, 3–5.

89 Ricoeur, *A Ricoeur Reader*, 427.

90 Ricoeur, *Time and Narrative, Volume 1*, 40.

91 Kearney, "Narrative Imagination: Between Ethics and Poetics," 173–90.

92 Ricoeur, *Oneself as Another*, 115.

4

Developers' Sphere of Deliberation

Spheres of Deliberation

The concept of narrative deliberation introduced in the previous chapter applies to several contexts of the Digital Poetics that we call spheres of deliberation. The spheres of deliberation represent prototypical positions of interaction with digital technologies in which narrative deliberation can contribute to decisions about ethical dilemmas related to the design, use, and institutionalization of new technologies. Although it is possible to think of the taxonomy of these spheres based on various criteria, this book addresses three spheres representing distinct forms of the relations between digital technologies and personal and institutional agency. These spheres intend to capture active and passive dimensions of human relationships with digital technologies. On the one hand, we have agency over what to develop, how to design it, and which values to prioritize. On the other, the development of digital technologies creates unforeseen personal and social impacts, and often, digital technologies can impose themselves as seductive, addictive, and dominating forces on how people interrelate, value their actions, and participate in social and political communities.

The first sphere of deliberation, which is explored in this chapter, is that of the developers of new digital technologies. This sphere investigates how narrative deliberation should lead creators of digital technologies to reflect on the products they conceive and on the impacts of these products on users. The second sphere of deliberation, which we study in chapter 5, deals with the end users of digital technologies. It investigates how these technologies impact the end user's agency and how narrative deliberation by the user community is essential for the new mixed forms of agency, by users as well as developers, that digital technologies provide so that these technologies may better align with users' plans for a good life. Finally, the third sphere of deliberation, analyzed in chapter 6, considers how social agents, public servants, and political representatives should deliberate regarding digital technologies to ensure these technologies foster mutual recognition and increase the possibilities of creating and solidifying fair institutions. Along the way, we will return to some of our analyses of Ricoeur but also consider some additional facets of his writing that are applicable in these situations of more specific application.

Wise Development

This first section introduces the chapter's topic and investigates how practical wisdom, informed by narrative deliberation, is essential for developing new digital technologies. The three subsequent sections turn to more specific contexts of application in this context. As mentioned in previous chapters, the acceleration of technological enablers' development, such as smartphones and artificial intelligence, and the tremendous penetration of social technologies in society—digital ubiquity—constantly create new ethical dilemmas that predetermined general rules and principles are insufficiently equipped to address and require additional recourse to practical wisdom.

The fundamental premise of the sphere of developers in Digital Poetics is that digital activity transforms individual and collective agency and, therefore, brings with it an intrinsic ethical dimension. Admittedly, there is something profoundly utopian about this proposal when we consider how current economic values dictate a vision of the development sphere limited to the creation and optimization of digital artifacts to maximize their functionality and economic returns. However, as we commented in the second chapter, in recent years, the need to think about developing new technologies within an expanded framework of Digital Poetics that integrates technical and ethical aspects has become more evident. Within this framework, good technological development is not only technically and scientifically competent but also ethically committed and consistent. Development teams must be both skillful (excellent at digital *poiesis*) and wise (excellent at ethical deliberation in their *praxis*).

Faced with the unprecedented ethical dilemmas that digital technologies create, it is impossible to apply general rules mechanically; they do not cover specific aspects of particular ethical issues. While rules and guidelines recommend what should clearly be avoided or promoted in developing new technologies, these rules are defined at a high level of abstraction in an effort to encompass multiple scenarios, often meaning they are not perfectly fit for deciding ethical dilemmas stemming from ever-changing digital innovations. For example, nearly all documents with ethical principles and guidelines for developing digital technologies recommend that consent precede user data collection. The issue lies in applying this principle to a particular artifact's specific context. For instance, mobile application developers utilizing users' geolocation data to offer a service need to decide when to collect the data, how to write the consent request message, how often to show this message, how to store the data consistently with the consent conditions, and so on. Practical wisdom in the context of digital technologies deals with applying principles to particular cases of artifact development. In chapter 3, we analyzed on a conceptual level how practical wisdom is an integral part of narrative deliberation. In this chapter, we revisit the concept of practical wisdom in the context of narrative deliberation, but with the specific aim of investigating its application in the sphere of digital artifact developers.

In *Critique and Conviction*, Ricoeur connects the concept of hermeneutic application and *phronesis*, emphasizing the significance of practical wisdom in his "Little Ethics." Ricoeur asks, "What does it mean to resolve a novel ethical-practical problem?"[1] He identifies this question as a problem of practical wisdom, which he aligns with the hermeneutics of "application" under the guidance of Aristotelian *phronesis*. Although Ricoeur directs his attention to challenging cases in law, medicine, and everyday life, his assertion clearly indicates the association between *phronesis* and ethical considerations surrounding emerging technologies. Digital technological advancements introduce unprecedented circumstances that demand phronetic reflection, which entails the application of ethical principles and regulations to specific situations through innovative ethical thinking.

The following sections explore three dimensions of practical wisdom that should underlie the developer's deliberative process so that ethical ends can effectively balance and shape the acceleration of technical means. First, in section 2, practical wisdom applies general moral principles to particular ethical situations that constantly require adjustments due to the speed of digital development. However, these adjustments must be guided by a sense of respect and concern for the users of new technologies. Thus, in this section, we propose that the development sphere of Digital Poetics should be marked by a relationship of solitude between developers and users. To this end, we will employ Ricoeur's threefold mimesis and his theory of the world of the text.

Next, in section 3, we endorse for the developer sphere a plural *phronimos*. As we explored extensively in the previous chapter, the narrative deliberation that applies practical wisdom is not the work of a single person. Given the technical and social complexities surrounding emerging technologies, it is unrealistic to think that a single person can deliberate well on all the ethical and technical aspects necessary for the responsible development of new functionalities. In the sphere of developers of Digital Poetics, practical wisdom emerges from a joint, interdisciplinary effort, and this is a plural *phronimos*.

Finally, in section 4, we argue that the creative imagination that is usually associated with technical advances should also be a feature of the ethical aspects of Digital Poetics in the developer sphere. Faced with a new ethical dilemma created by technical advances, general moral principles can be inadequate or conflicting. Practical wisdom involves an exercise in moral imagination to devise a particular new decision that meets the intentionality of the general principles as fully as possible but also adapts them to the demands of respect and requests from the various users of digital technologies.

Developer Solicitude of and Respect for Users

Responsible Developers

In *Oneself as Another*, Ricoeur states that practical wisdom involves the development of conduct that effectively addresses the exceptions that solicitude and respect for others require, while deviating from established rules only to the slightest degree necessary.[2] This section examines the relationship between developers and users based on the concepts of developer solicitude and respect.

The first conceptual difficulty to address is that Ricoeur organizes human relationships into three categories, starting with the relationship with oneself, which implies and expands to face-to-face relationships with others with whom we interact directly, and finally to other members of society with whom we have no direct personal contact.[3] In this tripartite framework, caring and respect for the other refer primarily to direct relationships between people in face-to-face contexts. This peculiarity raises the objection whether the concepts of solicitude and respect are appropriately applied to software developers as they rarely make decisions in the context of direct interaction with users. More broadly, such an objection questions the nature of the relationship between developers and users. We must clarify this relationship in order to proceed consistently with the proposed application of the concepts of solicitude and respect in the deliberation of the development of new technologies.

As discussed in chapter 2, given the evolution of software engineering toward a consumer-based industry mediated by virtual application stores, developers clearly cannot have a personal relationship with each of their users, at least in the case of most applications accessible to private users. But does this lack of a connection mean that we have to frame the developer and user relationship as completely segregated?

In the early 2000s, Gotterbarn criticized the framing of the relationship between developers and users that created a division between them.[4] Gotterbarn's main concern was the responsibility that should be attributed to developers when serious software problems lead to undesirable effects on users. As we analyzed in chapter 2, Gotterbarn argued that two critical factors created this valley that separates developers from users. The first is the mistaken view that software practice is ethically neutral. The second cause of the gap is the complex structure of software development, involving multiple actors and processes, which leads to a diffusion of responsibility.

Gotterbarn summarized his diagnosis of the problem by saying that "many problems in software development can be traced to a narrow understanding of professional responsibility."[5] In response, he offered the concept of positive responsibility, which "emphasizes the virtue of having or being obliged to have regard for the consequences of his or her actions on others."[6] Although Gotterbarn's concern at that seminal moment in software engineering practice was pragmatic, his observations already pointed to the need for a new framing of the developer–user relationship. The near disappearance of the end user from the developer's field of vision, as if the consequences of technology production were not a primary concern, leaves developers with a "narrow understanding" of responsibility. Setting aside until later in the chapter a deeper analysis of the problems of attributing responsibility in complex digital systems, another concern to retain from Gotterbarn's argument is that the concept of positive responsibility already points to the need for a new type of relationship in which the "other/user" returns to technology developers' field of vision.

The Model of the Text

A first obstacle to a framework promoting a conceptual and practical rapprochement between developers and users on an ethical level is the real impossibility for developers to relate directly to and consider the peculiarities of an artifact's millions of users. In fact, users will often have contact with the artifact only long after the developers have finished their deliberative and creative process. Therefore, it is necessary to seek another model for the developer–user relationship that escapes the pragmatist reductionism of market relations, recovers aspects of interpersonal responsibility, and addresses the opacity of each user from the perspective of developers.

If we take Ricoeur's hermeneutics of the text, considering authors and readers, as a model for comparing the relationship between developers and users, we gain a new, more appropriate perspective for the ethical discussion of software development. For Ricoeur, the world of the text mediates the relationship between authors and readers. In one article on this concept, he explains that "a literary text in general, a narrative one in particular, projects before itself a world-of-the-text, a possible world, to be sure, but a world nevertheless, a place I can think of myself inhabiting to carry out there my own-most possibilities."[7]

One of the fundamental ideas of this book is that the creation of digital devices should follow the poetic model of the configuration of meaning through cultural works. Thus, we should look at the design of digital artifacts as similar to writing a text to explore the limits of this comparison. The parallel between text and digital artifact is, in fact, an intuition that other philosophers of technology who study Ricoeur's work have already noticed and discussed.

David Kaplan was one of the first authors to recognize the potential of Ricoeur's philosophy for ethical issues in emerging technologies. In "Paul Ricoeur and the Philosophy of Technology,"[8] Kaplan suggests that some contemporary approaches to emerging technology studies emphasize the hermeneutic and contextual interpretation of technology as a social reality in relation to the cultural backdrop of meaning. From this perspective, Ricoeur's work provides a valuable philosophical framework for understanding the meaning and importance of technological practices. As we have in part already discussed, this framework includes Ricoeur's hermeneutic theory as a way of mediating the technical and social dimensions of technology, his narrative approach to illuminate how technology integrates with the stories of our lives, and his moral-political philosophy to assess the appropriateness and ethical considerations of technology. Kaplan argues that the concept of mediation, as synthesized in Ricoeur's hermeneutics, offers a means to interpret the diverse ways in which artifacts mediate experiences and a vehicle for responding to their limitations.[9]

In particular, for our discussion of the relationships between developers and users, Kaplan presents the Ricoeurian model of the text as a paradigm for mediating meaningful experiences, suggesting that technologies are similar to texts in that they are readable and gain meaning based on their context. Kaplan recalls that Don Ihde applied Ricoeur's notion of indirect and mediated experience to our experience with technology, highlighting how devices that are "read," such as telephones and step counters, display particularly well the hermeneutic character of technology.[10]

One of Kaplan's broader concerns is to show how the model of the text applied to technology escapes the deterministic paradigm of some philosophers of technology, where technology is assumed to entail a methodology of control.[11] Kaplan's approach leaves room for diverse encounters that technological configuration opens to the world and for the various forms of refiguration that users will make. Although this openness to different appropriations is also a constant concern in our discussion about the use of narrative imagination to explore diverse refigurations, in this section we want to maintain a focus on the possibility of applying Ricoeur's hermeneutic model, which is primarily grounded in written texts, to configuration through other media, such as digitally in the form of technological artifacts.

Ricoeur himself explores the applicability of the model of the text to other configurations of meanings in "The Model of the Text: Meaningful Action Considered as a Text."[12] In this piece, Ricoeur investigates to what extent the hermeneutic model based on texts written in natural language can be a productive paradigm for the object of the social sciences and to what degree his text interpretation methodology offers a viable model for interpretation in the field of the human sciences. Ricoeur proposes two essential criteria for expanding his semantic theory to the analysis of human actions. First, the object of analysis must have four essential characteristics of written texts, and second, the methodology used in the study must follow the model of text interpretation. We look at these requirements next and their possible application in the context of digital technology development.

Let's start with the four fundamental characteristics of written texts. The first is that they are discourses fixed and inscribed in a stable form, as opposed to the fluid and transient nature of the spoken event. Ricoeur uses Austin's and Searle's speech acts theory to clarify that three dimensions of meaning in the spoken discourse are fixed in the written text: the utterance (locutionary act), the practical intent of the speech (illocutionary act), and the effects or responses of the speech receiver (perlocutionary act).[13] Thus, when giving an order, "Open the door!" the written text captures the predicative utterance, the practical intention of being an order—typically denoted by grammatical devices such as tenses or punctuation marks—and eventually the effect as fear or reverence that such an order awakens in the interlocutor. Such translation, though, from spoken discourse to written text of the three dimensions of speech acts occurs in varying degrees. It is easier to capture the utterance that can be transferred from the verbal model of sounds to the model of written language given the typical correlation of verbal and written signs in natural languages. By contrast, the illocutionary and perlocutionary dimensions require other forms of translation into written language to account for the non-linguistic and contextual semantic aspects of the discourse event.[14]

With respect to this first criterion of a fixed and stably inscribed form, the development of digital artifacts, particularly those involving software-based user interfaces, follows the text model. Consider, for example, a screen that asks for a username and password. Through the programming language, the developer inscribes in the software, the statement of the request, and the grammar of the interface (which may involve specific graphic icons) capture the intent of the request.

Ricoeur explains that the second characteristic of the written text concerns a certain break between the intended meaning of the author's speech and those meanings the written text potentially evokes.[15] According to Ricoeur, when the text is written, the author's original intention and the text's meaning may no longer coincide. This separation between the textual meaning and the author's intention is the main issue in writing. For Ricoeur, it is not that we imagine a text without an author; the connection between the writer and the text remains, but it becomes more complex and enlarged. He summarizes this rupture by indicating that a "world" is opened up by the text: "What the text says now matters more than what the author meant to say, and every exegesis unfolds its procedures within the circumference of a meaning that has broken its moorings to the psychology of its author."[16] This criterion also applies to software development. A certain button's intended meaning on a digital interface can be interpreted in different ways by users who cannot directly ask the software author about the button's meaning. Even if textual meaning goes beyond an author's intent, though, it remains circumscribed by the semantic context. Similarly, for our argument, it is crucial that a circumscription of meaning still exists even if no longer directly linked to the author's discourse This circumscription defines the contours of fluid margins of meaning that readers with a certain degree of reasonableness can explore. Thus, even though there may be various ways to make sense of an application's login screen or other less common forms, the digital action leaves a demarcation of meaning that traces a region of possible interpretations that circumscribes user's choices.

The third characteristic of the written text is that it opens up a world of meanings that are no longer directly and ostensibly referenced through the author's speech situation but which are projected by the text itself: this is the world of the text. Ricoeur argues that we use, for example, the term "world" of Greece not to describe the specific experiences of those who lived at the time, but rather the enduring references that continue even after those circumstances disappear. These references now serve as possible ways of being and as symbolic dimensions of our existence.

This dimension also finds an analog in software development. When creating an application, the developer imagines the relationship between the user and a certain aspect of reality, such as requesting a ride. Then the developer inscribes this experience in the form of screens, interactions, and algorithmic decisions that refer to the phenomenon as experienced at the time of writing the code. After that, the code no longer makes direct reference to the experience that the developer had in mind; instead, it continues to open a horizon of meaning that represents the interactions that countless users can have with the reality mediated by the digital model.

Ricoeur culminates his inquiry into the distinctive attributes of the written text in a fourth characteristic that offers a reflection on the addressee of a communication. He acknowledges that communication is fundamentally based on speech, not just language, for speech is directed at an interlocutor. While in face-to-face communication, speech is oriented to a co-present interlocutor, written speech is directed at anyone who has the ability to read. As a result, in writing, the dialogic relationship becomes less tight, since the written discourse is not specifically aimed toward a single interlocutor or a co-located group of interlocutors but is offered to a self-created audience. Instead of being addressed only to the persons who share the context and the ostensive references, the written text addresses an audience that is normally dispersed both geographically and temporally.

Digital development seems to analogously follow this criterion of expansion and dispersion of the recipients of the author/developer's poetic creation. As we saw in the second chapter, the history of software engineering started with relationships that resembled face-to-face dialogue, because developers created unique and customized solutions for specific groups of users who often were coworkers in the same institution. However, with the internet and mobile devices, an expansion of the face-to-face relationship occurred, making the expression of the software model open to a self-created audience, as Ricoeur suggested in the model of the written text.

We have taken significant steps to show that Ricoeur's criteria for analyzing written texts apply analogously to the configuration of digital artifacts. We thereby move forward in exploring the poetic dimension of the development of digital technologies and in providing a more promising analytical framework for the study of ethical deliberation in the developers' sphere.

Implied Developers

We must address at least one further variable before we can consistently apply Ricoeurian textual semantics to the digital world. An apparent contradiction exists between the distance created by the world of the text that separates the author and reader of a written text and our attempt to bring developers and users closer together. If we want to continue to probe Ricoeur's insights in this context, it might seem at first glance that he has not explicitly thematized the ethical implications of the author in relation to the readers of his or her works. While in some of his work oriented to hermeneutics this may be true, we can find at least two relevant points in his later work where he explores the author's ethical responsibility toward the reader.

First, in *Memory, History, and Forgetting*, Ricoeur's analysis of the duty to retell certain historical episodes with the intention that they are not repeated shows that he did not ignore the ethical dimension of the relationship between authors and readers.[17] Further, across many pages of the third volume of *Time and Narrative*, Ricoeur discusses the dialectic between the world of the text and the world of the reader that clearly touches on several aspects of the possible ethical relationships between authors and readers.[18] This dialectic defines a spectrum of agency in the work of the text, from the near hegemony of power of an author who manipulates his or her readers to the opposite pole of readers' total interpretative freedom.

In *Time and Narrative*, Ricoeur begins his inquiry in dialogue with Wayne Booth, analyzing the agency of authors over readers through textual rhetoric.

Booth's *The Rhetoric of Fiction* focuses its attention on the initial pole of work of the text analytical spectrum and emphasizes the power of authors to persuade their audience.[19] For Booth, the "implied author" is the author's image or persona created through the narrative choices made in a work of fiction. It is not the same as the real-life author but rather an idealized and imagined version of the author who is responsible for the structure, style, and thematic content of the story. The implied author is a construct that helps the reader understand the narrative and the author's intentions behind it. Booth maintains that the narrative choices the implied author makes, such as point of view, structure, and style, can have significant moral implications. These choices influence how readers experience the story and engage with its characters, themes, and messages. Ricoeur too recognizes that "the theory of reading has warned us that the strategy of persuasion undertaken by the narrator is aimed at imposing on the reader a vision of the world that is *never ethically neutral*, but that rather implicitly or explicitly induces a new evaluation of the world and of the reader as well."[20] Ricoeur is, however, reluctant to fully accept the implied author's rhetorical impact as he wants to endorse the reader's role in the mimetic cycle. For Ricoeur, the domain of literature necessitates a reader capable of active engagement and responsive participation. This engagement exposes the limitations of an author-oriented rhetoric of fiction, as it relies solely upon the initiative of an author to transmit his or her personal viewpoint. The proposition that authors create their readers seems deficient in the lack of a dialectical balance between the active and passive dimensions of the act of reading.

Therefore, Ricoeur complements the rhetoric of fiction with an aesthetics of fiction in the broad sense of *aesthesis* that ancient Greek philosophers proposed. In this sense, the aesthetics of reading is a type of active affection that combines the act of reading with the reception of the text. Ricoeur summarizes this relationship by highlighting the combination of activity and passivity: "This being-affected has the noteworthy quality of combining in an experience of a particular type passivity and activity, which allows us to consider as the 'reception' of the text the very 'action' of reading it."[21] The poetics of the text is then grounded in a fragile and always in-progress balance between an "aesthetic" paradigm and a "rhetorical" one. This shift enables a more comprehensive exploration of the diverse ways in which a literary work affects its readers. The reader whom the work impacts exhibits a unique combination of passivity and activity within a distinctive experiential context. Being affected by and responding to the text are integral elements of reading it. Ricoeur proposes the concept of work (*oeuvre*) in that through active reading the text becomes a work, a "co-laboration" between the author and the reader.

In sum, Ricoeur does recognize the author's agency in the mimetic cycle, including the ethical dimensions of the author's role in relation to readers. Yet the French philosopher's emphases lie elsewhere. First, Ricoeur wants to endorse the reader's work and the opening of the world of the text to diverse appropriations on the basis of the reader's agency. Second, consistent with the prior subsection, the notion of the implied author benefits Ricoeur's hermeneutic strategy, because it analyzes the author's moral impact through his or her work—through the world of the text—and avoids passage to the psychological intentions of the flesh-and-blood person with quill in hand.

Second, returning to the context of Digital Poetics, the question is whether the limits imposed by Ricoeur that circumscribe the relationship between authors and readers of written texts (particularly fictional ones) find analogous traits in the relationship between developers and users. It seems reasonable to suggest that users can experience digital artifacts in their own right without the need for speculation about the possible psychological states of their creators. Digital artifacts also open up a digital world, allowing immanent explanations, the assumption of an "implied developer," and the distancing of the analytical process from the real-world developer. It is also essential to recognize the agency of users who appropriate digital artifacts in different ways and in various social, economic, and cultural contexts. Such a perspective is fundamental to avoid the fallacy of technological determinism that ignores users' agency. However, I believe that, as in literary mimesis, digital mimesis does not exclude the author's agency and, therefore, the developer's responsibility concerning the configurative action because—as in fictional texts—there is an immanent rhetoric that informs the digital world and impacts the possibilities of refigurations and the work of readers and users. The real-world author is responsible for the implied author, even if a hermeneutic approach does not require an ethical assessment of the former.

In addition to these considerations related to the world of text, there are characteristics specific to the configuration of the digital world that accentuate the ethical dimension of the relationship between developers and users. To explore these peculiarities of the digital configuration, let me propose a few conceptual clarifications that will pave the way for the remaining steps of our analysis in this chapter. The written text creates a spatial and temporal gap between author and reader, just as the design of digital artifacts creates a spatial and temporal gap between the moments of design/implementation and the moment of artifact use by users. The textual and digital worlds mediate this gap, respectively, between author and reader and between developer and user. Such mediation represents the possibilities opened up by the configuration of text and digital technologies. Although it is impossible to limit the interpretative and transformative possibilities that these worlds open, the possibilities are not infinite. The tension between the horizon of expectation of the text or digital artifact and the potential interpretations of readers or users reflects the malleability of the worlds that literary and digital configurations propose.

The Digital Mimetic Cycle

Thus far, we have made some progress in our attempt to clarify the relationship between developers and users using the paradigm of the relationship between authors and readers drawn from Ricoeur's textual hermeneutics. Our aim is to create a conceptual framework that takes account of the distance that the mediation of technological artifacts creates between developers and users, but that also restores the relationship of responsibility for the creative actions of developers in relation to the impacts they create on users. In this way, we can ground the narrative deliberation of the developer sphere on the basis of solicitude and respect for users.

The Ricoeurian paradigm of textual hermeneutics has allowed us, on the one hand, to consider the digital world created by developers as analogous to the world of text created by writers. Ricoeurian textual hermeneutics has also permitted us to think about the agency of users who can resignify the technological configurations of the digital world because, like texts, digital configurations "resembles a musical score lending itself to different realizations."[22] Yet, on the other hand, Ricoeur's emphasis on the active role of the reader seems to undermine a possible ethical attribution to authors of responsibility for the creation of meaning. This difficulty seems linked to the difference in the types of mediation between the textual and digital worlds. In order to unravel this difference and find a productive way around it, we need to take a closer look at Ricoeur's semantic theory through a new glance at the threefold mimesis.

As we discussed in chapter 3, the threefold mimesis is composed first of prefiguration (mimesis 1) that captures the way human action is inextricably symbolically mediated, with meaning inherently incorporated into action. In turn, configuration (mimesis 2) captures the author's creation of the world of the text in the inscription of the text in written form. And refiguration (mimesis 3) captures the possible adjustments of meaning precipitated by the reader's encounter with the world of the text in the act of reading. Before applying the threefold mimesis to digital technologies, it is productive to probe the extent to which Ricoeur's semantic theory can be consistently applied to other meaning-creating contexts so that we can validate our approach.

Ricoeur himself applied the threefold mimesis model developed in *Time and Narrative* to another form of expression in at least one of his articles. In "Architecture and Narrativity,"[23] Ricoeur proposed a close relationship between architecture and narrativity, as architecture configures space just as narrative configures time. There is a correspondence between construction and edification in space, on the one hand, and the creation of plots and stories in time, on the other. Ricoeur explores the parallels between architecture and narrativity throughout the text using the three mimetic moments as a guideline. He describes the process starting in a prefiguration stage that is linked to the idea of inhabiting. In a second stage, architects intervene in the preexisting inhabited spaces through building. Finally, there is a third stage of refiguration where architecture rereads our cities and dwelling places. One of the exciting aspects of this application involves the transition from the written media of literature to the media of architectural building materials, as this reveals a more general and comprehensive aspect of Ricoeur's semantic theory and suggests we may properly attempt to rethink it in the context of digital media.

Digital technologies also offer a way of configuring digital space and the experience of time, and this seems promising for our attempt to extend the application of the Ricoeurian threefold mimesis to Digital Poetics. In "Architecture and Narrativity," Ricoeur focuses on the parallel between the configuration of time—characteristic of narratives—and of space—characteristic of architecture. However, the expansion of the field of application of Ricoeur's semantic theory gains strength when we look at the threefold mimesis not only as a cycle of configuration and refiguration of time and space but also of meanings. Ricoeur pointed to semantic innovation as a common denominator between his study of metaphors and narratives and explicitly spoke of the configuration of meaning that emerges through language by means of the productive imagination. Semantic innovation can be extended not only to language but to all other domains that are symbolically mediated, as they have a linguistic dimension that permits linguistic transformation. Ricoeur's semantic theory, then, can be extended to other poetic activities of configuration and refiguration of meaning. This supports our endeavor of expanding the application of his semantic theory to Digital Poetics as an activity that produces and reshapes meanings.

I am not alone in recognizing digital technologies as an area where Ricoeur's broader semantic theory can be fruitfully deployed. Alberto Romele has already explored this potential application of Ricoeur's philosophy in his effort to address issues related to the specific context of digital technologies through the use of Ricoeur's hermeneutics.[24] Romele proposed reframing and expanding Ricoeur's hermeneutics to better address the materiality proper to digital technologies, as we discussed in chapter 1. He affirmed that while Ricoeur's approach differs from other ontological hermeneutics by placing greater emphasis on externalizing and materializing language and interpretational processes, the philosopher's view of materialized language is paradoxically idealistic. Ricoeur focuses on the underlying semiotic and semantic rules governing signs, symbols, linguistic metaphors, and narratives, fundamental for conveying meaning. However, he does not delve into the physical materialities that facilitate the transmission of meaning through these elements. Romele then went on to explore how a hermeneutic approach could consider the specificities of digital materiality.

Also drawing upon Ricoeur, Cockelbergh and Reijers, as mentioned in chapter 2, Cockelbergh and Reijers developed the concept of narrative technologies to analyze the temporal dimensions of technologies, highlight their social and linguistic aspects, and re-elaborate their mediating role in relation to their narrative capacities.[25] They argued that, like texts, digital technologies configure events and organize the structure of meanings of human action and time.[26] Cockelbergh and Reijers pursued their arguments based on two hermeneutical distinctions arising from Ricoeur's interpretative model of texts: between active and passive narrative technologies and between engaging and abstract narrative technologies. Regarding the first distinction, they analyzed the similarities between technologies and the paradigm of the written text paradigm, examining how technologies actively configure characters and events into a meaningful whole (instantiating a plot process).[27] In this way, they proposed to analyze how technologies' linguistic and material aspects interrelate. Concerning the second distinction of narrative technologies, Cockelbergh and Reijers considered the degree to which technologies configure narrative structures that are abstracted from the world of action. Their framework further corroborates with the theoretical approximation proposed in this section between developers and authors and users and readers. Their analysis highlights technologies' social and linguistic aspects and re-elaborates their mediating role, emphasizing their narrative capabilities and active shaping of characters and events into a meaningful whole. As a corollary, the narrative technologies approach brings to the fore the responsibility of the developer's choices in creating digital "emplotments" and their relationships to users' agency.

Based on this conceptual examination of the possibilities of applying Ricoeur's semantic theory to other meaning-making practices, we can revisit the mimetic moments within the context of the development of digital technologies. Software developers inhabit a world of meanings (prefiguration) and engender and inscribe new meanings in digital artifacts, creating a digital world (configuration) that finds and precipitates changes (refiguration) in users of digital technologies. We can then speak of a threefold digital mimesis and ask ourselves about its specificities in relation to the threefold mimesis of the written text as Ricoeur proposed. This will lead us back to our central question of this section concerning the ethical relationship between technology developers and users.

There are two salient differences between digital and textual mimesis: the narrowing of the refiguration space and the iterativeness of the configurative moment. We will discuss each as both are key elements of digital developers' ethical responsibility. The narrowing of the refiguration space means that the boundaries found in the digital world are more rigid than those of the world of the text. Typically, when reading a book, readers have a larger refigurative space because they bring so much more of their prefigurations into their interpretations of the world of the text. This refigurative space is correlated to what Ricoeur refers as the gap between "the horizons of expectation of the text and those of the reader."[28] For example, the textual description of a placid lake surrounded by dense vegetation on a pleasant autumn morning requires a substantial imaginative complement from the reader so that the possibilities of this spatial description gain a poetic referent. The writer's lake will almost certainly be very different from the lakes of all readers, yet this is one of the riches of the mimetic process and marks the plasticity of the world of the text. By contrast, the textual description of the lake, when configured through the model of other media, such as cinema, adds elements that inhibit the audience's refigurative scope. When filming a certain lake in a certain region with specific shoreline vegetation at a certain time of the year, the cinematographic producer narrows the refiguration space, because the film has a greater configurative vehemence.

Similarly, the configurative vehemence of the digital medium also creates a narrowing of digital refiguration. Developers not only describe a certain interaction that users can have with the digital world, but they implement this interaction in a way that restricts users' interpretative and creative possibilities when appropriating the digital world. This restriction does not entail a determinism of the digital configuration because the same communication application can be used for communication between friends, for requesting professional services, and even for users' appropriations that developers had never imagined. However, digital media offers a more active space for symbolic representation and, thus, restricts the user's agency in the process of interpreting the digital world. Within the Ricoeurian framework, we could say that in contrast to the writer, the rhetoric of the developer has a greater potential to persuade due to digital materiality. Let me give two examples to illustrate this point.

When creating a digital application, developers configure a technological narrative, as Coeckelbergh and Reijers proposed, which organizes time in a certain plot of events that have users as quasi-characters who begin to reshape their experience on the basis of the digital application.[29] Thus, when someone uses an application to order a pizza delivery, the application configures the moments of the pizza order in a temporal sequence that constrains in many ways how the user experiences this activity. Given the ubiquity and convenience of digital devices, this digital configuration of the pizza-ordering experience could become the prevalent meaning of "ordering pizza."

Messaging applications configure the communication experience through text editing widgets, buttons for sending messages, and indications that the message has been seen, opened, and, in some cases, that the interlocutor is writing a reply. Developers configure the time and experience of exchanging messages, and then users, when interacting with digital technology, refigure their experience and understanding of what it is to chat with another person. In many cases, messaging application features allow changes in the communication flow. There are also settings that increase users' agency and allow them to determine how they want to experience the exchange of messages. However, the configuration that developers propose impacts the users' possibilities of experiencing a chat. This impact on the users' meaning and agency serves, I argue, as an index of ethical responsibility that the relationship between developers and users establishes and that the digital world mediates.

The second difference between textual and digital mimesis involves the iterativeness of the configurative moment. Authors typically only have one substantial iteration in the setup cycle of a literary work. Though the possibility of versions and revisions exists, these are rare and usually do not change essential aspects of the original configuration. Thus, the interaction mediated by the world of the text is typically unidirectional and singular. By comparison, the configuration of the digital world is much more flexible. The instant global distribution and functional flexibility of digital technologies allow constant updates of current digital artifacts and architectures. User feedback often guides these updates, which can be made to decrease the refigurative space that may be creating undesirable practical implications.

What are the implications for the relationship between developers and users of this analogy between digital and textual mimesis? First, the cycle of digital mimesis clarifies how digital configuration can impact users' refigurations of meaning. The different degrees of configurative vehemence generate a substantial distinction between the model of textual narratives and that of digital narratives. Ricoeur argued: "What the strategy of persuasion [of textual narratives], wrought by the implied author, seeks to impose on the reader is, precisely, the force of conviction—the illocutionary force, we might say in the vocabulary of speech-act theory—that upholds the narrator's vision of the world."[30] In the digital configuration, developers have greater power to impose the force of conviction, and they do so in part through limiting the space for refiguration opened up by digital technologies. Further, as we discussed in chapter 1, the digital ubiquity that creates social expectations also makes it much more difficult for users to decide to distance themselves from certain digital configurations. This reduces their freedom to reject the resignifications of experiences proposed by the digital world. The increase in power of the rhetoric of developers correlates directly with a decrease in users' refigurative freedom.

The vehemence and iterative nature of the configuration of the digital world amplifies the developers' power to act on users and the need for developers to apply narrative deliberation. Configurative vehemence allows developers to control the affordances of digital artifacts, significantly limiting users' forms of appropriation. For example, simple checks on which characters can be used in edit fields can prevent users with names in non-Latin alphabets from consistently entering their data.[31] On top of the configurative vehemence, the possibility of remotely monitoring the types of refigurations that users make increases the configuration power of developers who can change the affordances of digital artifacts in response to unexpected and unwanted refigurations. This peculiar feature of digital materiality increases the agency of developers in relation to users and makes them more responsible for ensuring that, even if previous versions of functionalities have created undesirable consequences, they can dynamically adjust them to ensure that certain types of interaction with users are promoted.

The application of the text model to the development of new technologies allows us to re-establish the relationship between developers and users based on the mediation of the digital world. While digital mediation leaves room for different appropriations by users, greater developer power demands of developer rhetoric a more responsible role. This framework attends to the nature of the digital world's mediation, but it also urges developers to consider the ethical implications of the various choices they make when configuring the digital world. It proposes a more balanced and realistic model of the developer–user relationship that escapes the dichotomy between the extremes of either total isolation or a fictitious face-to-face. By considering developer responsibility, my framework proposes a theoretical foundation that seeks to explore a middle ground for developers between claims of technological neutrality or of determinism.

The threefold digital mimesis contests technological determinism, since the digital world developers configure a proposal of meaning that users may or may not accept. The developer's proposal of meaning is always part of an existential movement of personal refiguration that is open to users' choice. On the other hand, the configurative vehemence of the digital mimesis denies a false idea of technological neutrality because in fact the configurative act of designing digital artifacts imposes actual limits and suggests specific forms of refiguration. These forms affect the agency of users who refigure their experiences on the basis of the mediation of specific technologies. For example, users can limit the impact of social networks by restricting how they interact with them, but each social network application has characteristics that restrict how such refigurations can occur. In some cases, it is possible to choose not to see certain types of posts, but limiting infinite scrolling or customizing how posts are ordered in the timeline is impossible. We will explore this middle ground between technology determinism and neutrality further in the following sections.

It is precisely because of the digital world's vehement configuration that the ethical aspects of the relationship between developers and users must always be on the developers' horizon of concern. Developers are not just creating an ethically neutral digital world. They are constantly creating a world where users will potentially inhabit and reshape their experiences and, therefore, will have their agency impacted—certainly in different ways and degrees, but even so, limited to some refigurative possibilities delimited by the configuration of the digital world.

Ricoeur highlights the idea of "power over" the other that demands ethical consideration in the relationship between agents and recipients: "What must first be taken into account is a situation where one exercises power over the other, and where consequently to the agent corresponds a [recipient] who is potentially the victim of the first's actions."[32] The relationship between developers and users seen from the digital mimetic cycle highlights the developer's power over users mediated by the digital world. Such mediation makes explicit consideration of developers' ethical responsibility intrinsic to the relationship with users.

In the following section, we explore some of the features of narrative deliberation that are particularly relevant for the developer's sphere given the type of relationship between developers and users that we have derived from Ricoeur's semantic theory.

Synchronous Plural Phronesis

In chapter 3, we developed the concept of phronetic plurality from a narrative approach to practical deliberation. Employing this concept here, we expand Ricoeur's proposal on the possibility and importance of considering the "collective phronimos," a group of people who evaluate in a collective deliberation process a decision to be taken. Ricoeur summarizes his view on this possibility: "moral judgment in situation is all the less arbitrary as the decision maker—whether or not in the position of legislator—has taken the counsel of men and women reputed to be the most competent and the wisest. The conviction that seals the decision then benefits from the plural character of the debate. The phronimos is not necessarily one individual alone."[33]

Ricoeur's observation occurs in the ninth study of *Oneself as Another* in the context of the rapprochement he makes between practical wisdom and the Hegelian *Sittlichkeit* (situated morality). In particular, Ricoeur argues that one should not regard *Sittlichkeit* as a new instance above ethics and morals but rather as another way *phronesis* can be recognized in its application in the public and institutional domain. For Ricoeur, the public debate that leads to phronetic deliberation is a common feature of good ethical decisions. A plurality of people interact, mediated by language and culture, to carry out the narrative deliberation that is proposed in this book. In "*Éthique et Morale*," a text that prepares the basic lines of argumentation for his "Little Ethics," Ricoeur discusses the importance of institutional public debate for an inclusive phronetic deliberation that attends to the peculiarities of the case.[34] Ricoeur claims that public debate can embody what Antigone's chorus exalts as "good advice." The plural practical wisdom that emerges from the debate transcends individual concerns, manifesting itself as a shared, public *phronesis*, much like the debate itself.

The concept of plural *phronimos* is particularly appropriate when considering real situations of ethical decision-making in the development of digital technologies. These technologies carry social repercussions and complex technical characteristics that require a deliberative process that transcends the capacity of a single person. Deliberative narratives realize the necessary plurality of practical wisdom, so that developers configure digital worlds that are respectful of their users. These narratives should be the outcome of a genuinely open debate involving a diverse group of representatives pertaining to requirements, design and implementation documents, and coding. This discussion transforms the practice of technology development into Digital Poetics that simultaneously configures algorithms, screens, communication protocols, and ethical goals. Significantly, although our focus in this chapter is on developers, the plural *phronimos* in developer decision-making should come from a broader community than just developers themselves.

Consider, for instance, the development of mobile applications for tracking contagious diseases, as happened during the COVID pandemic. Such technological developments require multiple voices in the deliberative process. It is implausible that a single deliberator would know all aspects of the situation needed for phronetic deliberation. A phronetic plurality emerging from an inclusive debate during the requirements and design phase of development would involve representatives of the affected communities, legislators, software developers, mobile communication specialists, digital security specialists, infectious disease experts, and public health specialists. The deliberative narrative of Digital Poetics expands the scope of analysis so that developers may better act responsibly in their digital configurations. In so doing, we also expand the concept of developer to include non-technical participants in the process of designing and implementing digital technologies.

This deliberative process must create an integrated perspective from different viewpoints. The decision must betray as little as possible established moral principles—such as deontological codes of conduct—while maintaining the ideal of the good life in common. As we have highlighted, fast-paced advances in digital technologies make this decision even more complex and require immediate involvement of all parties in detailing the decision. Such a decision is not only proposed through language but is implemented through algorithms and computational systems, creating two layers of Digital Poetics that bind together technical affordances and constraints with social and ethical goals.

Recently, the principle that Ricoeur advanced on the importance of public debate in complex decisions has been recognized and integrated into several guidelines for developing digital technologies. Even if these guidelines make no direct reference to Ricoeur, his analysis helps to ground and amplify the merits of this approach. For example, the Berkman Klein Center's Principled AI document sets out a specific principle for developing digital technologies in an inclusive way that involves a diverse group of stakeholders.[35] The Center's principle of "Inclusive Design" underlines the fundamental notion that the responsible design of applications based on artificial intelligence (and we can add digital technologies in general) requires in the process a broader and more diversified involvement of requirements specification and projects. This principle manifests in two ways. Primarily, it emphasizes the need for diversity within AI design teams, encapsulating multiple perspectives that mirror society's diversity. A more developed example of this perspective is evident in the European Commission document "AI for Europe" which insists on incorporating into AI development a more significant proportion of women, individuals from diverse cultural backgrounds, and people with disabilities.[36]

This inclusive participation, starting with comprehensive AI education and training, is crucial to ensuring AI's non-discriminatory and comprehensive nature.

According to the Berkman Klein Center, the second dimension of "Inclusive Design" posits that a broad segment of society should be able to contribute to the discussion not only about AI project requirements but also about AI applications and contexts. Specifically, it suggests creating "a truly diverse and inclusive social forum for discussion to democratically determine which forms of AI are right for our society."[37] The Toronto Declaration echoes this perspective, stressing the need to involve end users in decision-making about the design and implementation of AI, in order to "ensure that systems are created and used in ways that respect the rights—especially the rights of marginalized groups who are susceptible to discrimination."[38]

The "Inclusive Design" principle requires developers of new technologies to employ a proper process for deliberation. The model of narrative deliberation offers a considered response to this practical need. The various interlocutors participating in deliberation on new technology creation bring to bear both their retrospective narratives (how they experimented with similar technologies in their living contexts) and their prospective narratives (how they imagine the impact of a certain technology in their living contexts). These various narratives are then considered when choosing the best ways to implement digital artifacts while respecting established moral principles, the idiosyncrasies of the decision at hand, and the ultimate goal of creating a digital configuration that aims at users' fulfilled lives.

The digital plasticity highlighted in the first chapter allows decisions to be expressed (configured) in the form of algorithms, communication protocols, and data structures. Digital Poetics manifests its full relevance in its possibility of changing design and implementation requirements and parameters based on decisions emerging from narrative deliberation that express a plural practical wisdom informed by encompassing moral principles.

An example of digital technologies research that considers the analysis of multiple personal and cultural values as fundamental to the development process is "A value-oriented and culturally informed approach to the design of interactive systems,"[39] which aims to "sensitize and support Computer Science and Engineering professionals in taking values and culture into consideration throughout the design of interactive systems."[40] The interdisciplinary approach this research group proposed provides a well-organized collection of tools and methods designed to help at every step of the design process. These tools and methods range from identifying the stakeholders and their values to organizing the requirements and evaluating the final solution, considering important social and ethical factors. Well-thought-out approaches like these address the fragility in recognizing all scenarios and the infinite application possibilities of a given digital technology. These approaches point to the viability of integrating ethical and social aspects into the deliberation of developers during the conceptualization, design, execution, and testing of digital technologies.

Reducing Unintended Consequences through Imaginative Variations

As we have seen in the threefold digital mimesis, the moment of refixation is marked by the rhetoric of the user who interprets and applies technological affordances in various ways, often with consequences that the configuration of the developers did not foresee or intend. This is another point at which narrative deliberation offers a productive framework for Digital Poetics to explore innovative alternatives to reduce unintended consequences in the development process.

In addition to malfunctioning algorithms, protocols, data structures, and security mechanisms, different users use digital artifacts in different ways. As we noted in chapter 1, Don Ihde has suggested that technologies are multistable in that they can have multiple potential roles, impacts, or manifestations depending on the context of their use.[41] Such multistability challenges more deterministic views of technology that suggest technologies have an inherent, unchanging nature that drives social change in a specific direction. Instead, according to the concept of multistability, the influence and effects of technology significantly depend on how they are incorporated and used in human practices.

Developers configure meaningful technologies that create a digital world open to new appropriations. In Digital Poetics, new meanings emerge in multiple ways from the encounter between users' prefigurations and the digital worlds configured by developers. The meaning of digital technologies is only fully realized in the context of use, making the problem of unintended consequences very complex.

Peter-Paul Verbeek agrees with Ihde and has affirmed that digital technologies "have no fixed identity but get defined only in their context of use."[42] Furthermore, he has maintained, "Technologies have to be interpreted and appropriated by their users to be more than just objects lying around."[43] Again, this argument is a refutation of technological determinism. Veerbek's point emphasizes the active role of users who appropriate the affordances technological artifacts make available in contexts of use and within different lifeworlds. In the case of digital technologies, at least two peculiarities impact Verbeek's assessment, which was intended for all types of technological innovation. First, the range of possible appropriations is even more remarkable given the pervasiveness of digital artifacts across the distribution networks we discussed in the first chapter. Second, artifacts that contain enablers with artificial intelligence have greater agency than other types of analog or even digital technological artifacts without artificial intelligence components. This increased agency of digital technologies can be a way for the artifact to adapt to the idiosyncrasies of each user. Think, for example, of the options video and music streaming applications offer.

However, the malleability of digital artifacts must not lead to the conclusion that we are back to square one and that a developer's ethical responsibility dissolves in the vast space of users' possible appropriations. On the contrary, in light of the multistability of digital artifacts and their role in mediating experience, Verbeek has postulated the need for a process of "mediation analysis," which can occur in two complementary ways:

> The first is simply a prediction by the imagination of the designer. Imagination can create a link between the designer and the technology in its future use context. . . . A second way to formulate an informed prediction of the future mediating role of technologies is a more systematic one. It consists in an augmentation of the existing design methodology of constructive

> technology assessment in such a way that it becomes an instrument for a democratically organized moralization of technology.[44]

Verbeek's analysis of technological mediations is insightful for our argument, as it emphasizes two fundamental aspects of a narrative deliberation approach to developing new technologies. First, it highlights the role of imaginative processes in exploring the unintended consequences of new technologies. For example, Verbeek cites a deliberation process focused on developing sustainable products. This process aims to extend the lifespan of products, thereby minimizing the impact of technological disposal. Second, Verbeek's insights underscore the need for collective deliberation from multiple perspectives. This multiplicity must involve other social actors who participate in the technologies' contexts of use. Regarding this external dimension, Verbeek refers to the method of constructive technology assessment (CTA), as Jacobson, Schot, Misa, and Rip describe.[45] CTA proposes a process that involves "feedback assessments of the technology-in-design by all relevant actors, such as users, pressure groups, designers, companies, and so forth, into the design process." This is accomplished by organizing meetings with all relevant parties, aiming to reach consensus on the design of the technology undergoing constructive assessment.

We will explore how to integrate narrative deliberation productively on the basis of Verbeek's two suggestions. The phronetic plurality discussed earlier is undoubtedly an essential component for achieving this objective. A multiplicity of voices in the deliberative process tends to bring up cases that a smaller, homogeneous group of software engineers might overlook. However, given the global scope and potential long-term impacts of digital technologies, it is necessary to explore other methods for analyzing their impact. This expanded impact brings us to Verbeek's concern about analyzing technological mediations. Although Verbeek well recognizes the need to use imagination to explore the different contexts of technology, it is fundamental to propose practical ways that such exploration can happen in a comprehensive manner and especially to take into account the forms of mediation that more fragile users will have with digital technologies. In this sense, the imaginative variations we propose as part of narrative deliberations emerging from fictional, historical, or journalistic narratives involving digital technologies will enrich and expand requirement and design considerations for digital products and reduce unintended consequences in early design stages of technological artifacts.

The concept of narrative deliberation can also structure the constructive technology assessment of CTA meetings and ground conceptually the expectations, roles, and processes of an extended group of people thinking about the possible forms of mediation of digital technologies. Recognizing the importance of mitigating the risks of unintended consequences and existing suggestions for processes to increase inclusivity in design, narrative deliberation offers a comprehensive and coherent conceptual framework to support initiatives that integrate ethical concerns into the specification processes and the design of digital technologies.

The training of future digital technology developers is another crucial application of imaginative variations in the sphere of developers' deliberation. Some researchers in U.S. higher education institutions have explored the application of fictional narratives to arouse an ethical sensibility of technological issues. Emanuelle Burton, Judy Goldsmith, and Nicholas Mattei have taught for several years a course in Science Fiction and Computer Ethics. They argue that "fictional [narratives] offer students a way to engage with ethical questions that cultivate their capacity for moral imagination; science fiction, in particular, can make the ethical stakes of blue-sky projects vivid, pressing, and immediate."[46]

In a paper aimed specifically at discussions of artificial intelligence within computer science studies, Burton, Goldsmith, and others have suggested that narratives, including books, plays, and poems, have always shown us our world.[47] They argue that narratives help future developers understand how life would look when their technological creations are deployed and how choices influence the future. The authors propose integrating discussion and applying ethical frameworks and case studies into the technical discussion in computer science and other digital technology-related academic programs. The simple exercise of narrative imagination to awaken a moral sensitivity in students already adds value to their formative process, even if the formal evaluative stage is not fully explored based on a set of formal concepts of philosophical ethics. The advantage of an approach that brackets the assessment stage is that computer science instructors can feel more comfortable adding small modules to their study plans without worrying about answering often complex questions about ethical frameworks. Even absent an ethical verdict, exposure to narratives that contextualize the impacts of technical concepts has the value of awakening students' concern for these issues that future courses on professional ethics or philosophy can discuss based on evaluative theories. This sensitivity will also be fundamental for future interdisciplinary discussions on how graduates will become participants in the inclusive debates we have proposed in previous sections.

This imaginative exploration is even more critical in cases of seminal technologies where there is an absence of widely distributed devices and impacts are uncertain. Nanotechnologies offer an existing example, but the issue is readily applied also in the context of emerging digital technologies. The scenario of new technology is probably the most significant, because ethical discussions must occur before the technologies are integrated into the lives of people and societies. Using the example of nanotechnology, Berne and Schummer have proposed the use of science fiction for engineering studies: "Through science fiction, engineering students are given opportunities to move beyond ideas of present reality material into the domains of the imagined future, where they can work with moral questions of our future with nanotechnology in creative and active ways."[48]

Overcoming Ethical Aporias with Narrative Imagination

One of the most essential characteristics of practical wisdom is the consideration of each particular case, whether existing or possible, in order to apply and adjust principles and rules to the idiosyncrasies of the specific situation. However, as we discussed when proposing the narrative deliberation process, phronetic deliberation often must resolve a moral dilemma in which valid general concepts contradict each other when applied to a particular case or consider resolution when no moral concept is adequate to the situation at hand. In both cases, phronetic deliberation must construct an ethical decision that remains as faithful as possible to general principles but also respects the rights of each person involved in the situation.

John Wall has advocated for an approximation between practical wisdom and *poiesis* that considers the poetic—the creative—potential of ethical decisions. As we discussed in chapter 2, Wall challenges the traditional separation of moral and non-moral activities, arguing that practical wisdom (*phronesis*) and poetics are intimately connected and mutually reinforcing. He argues that the break of creative expression from ethical discourse is a modern phenomenon and that, historically, creative settings such as narratives have often expressed moral life. Wall concludes his argument with some ideas that are particularly pertinent to the practice of software engineering:

> Although poetics has various dimensions in human affairs—artistic creation, technological advance, scientific discovery, cultural expression—it is also deeply involved, perhaps not so surprisingly after all, in moral activity as well. Indeed, poetics is arguably more radically involved in moral life than anywhere else, for here the process of creation is not just by but also between irreducibly other selves, and precisely in their very irreducible otherness.[49]

Three aspects of Wall's conclusion reinforce our argument. First, Wall explicitly casts technological advances and scientific discoveries as poetic expressions; second—supporting our fundamental hypothesis—he reaffirms the interweaving of ethical and poetic aspects; and finally, he highlights the need to include the "other" in this poetic-ethical process.

I want simply to extend Wall's discussion of his final point and then gather some final remarks for this section and chapter. In his evaluation of other contemporary thinkers who have reframed the Aristotelian concept of *phronesis*, Wall addresses Ricoeur's proposal, emphasizing the importance that embracing consideration of others has in the idea of critical *phronesis* that Ricoeur proposes in his "Little Ethics": "The purpose of Ricoeur's critical *phronesis* is not to resolve or sublate moral differences in some third ethical totality. Rather, it is to engage in the unending task of responding to others through an ever more radically inclusive moral narration."[50]

In the developer stage of digital technologies, the creative process of phronetic deliberation becomes a crucial step in inclusive requirements specification and design discussions. Such ethic-poetic creativity may be one of the more evident manifestations of Digital Poetics through the constant back-and-forth of phronetic deliberations and technical implementation decisions. Developers can creatively resolve a dilemma between data privacy and user agency values with changes in the artifact design, such as by changing default values, adding configuration parameters, and adjusting the experience based on user vulnerability (e.g., age). This deliberative process turns ethical decisions in technical implementations and digital technologies into artifacts steeped in ethics. The good news also is that deployment architectures and software plasticity make digital configurations far more fluid than other forms of technological configurations involving hardware. An issue with a car's electronic injection software can be resolved by deploying a new software version over the air. In contrast, a problem with the vehicle's seat tilt adjustment lever necessitates a recall, causing a substantial financial hit to manufacturers and considerable inconvenience for users.

In concluding these reflections on the developers' sphere of deliberation, we should recognize two potential limitations in this sphere that point to the need for a more comprehensive ethical-deliberative process, which will be the subject of subsequent chapters. First, while we have proposed that developers' narrative deliberation should presuppose respect and concern for users as its primary objective, this goal can be negatively affected by other conflicting objectives, such as a company's accumulation of economic value. These potentially misaligned goals are often determined by decisions of executive boards and propagated downward in corporate structures until they affect developers' goals. In response, developers' narrative deliberation can strive to maximize the implementation of technologies that enable companies to achieve sustainability and economic improvement while maintaining respect and concern for end users as primary objectives of narrative deliberation. However, breaking points in development may occur where respect for users can be overshadowed by corporate economic priorities. Faced with this possibility, developers' narrative deliberation needs to be balanced by a sphere of institutions that reinforce care for users and mutual recognition as the primary objectives of Digital Poetics. We will address this sphere in chapter 6.

A second limitation of the developer sphere is that even when it explores phronetic plurality and employs narrative imagination, the ubiquity of digital technologies makes it practically impossible to control all unintended consequences. Some have argued that, in this light, devoting priority to developer phronetic deliberation is misguided. Under this view, even aside from economic incentives, it is better to prioritize deliberation to occur after deployment rather than before. The flexibility of digital deployment allows for quick development and distribution to users of an initial artifact, and this flexibility permits quick refinements based on user feedback. Yet while this strategy has a role to play, it is not ethically sufficient. Some unintended consequences can be prevented through developer deliberation, and this is particularly pertinent in cases involving products that turn out to have a high social impact (which itself may be unknown without initial deliberation). The deploy now and fix later approach can create a scenario of negative effects that is especially problematic when involving a large and potentially vulnerable population, such as children and adolescents. As we highlighted in chapter 1, developers can monitor in real-time how users are using the functionalities of digital technologies that are always connected. In addition, developers can offer ways that are integrated into digital artifacts for users to provide feedback and suggest changes. It remains the case, therefore, that a defining trait of responsible developers in Digital Poetics includes the reduction of unintended consequences through a productive narrative imagination informed by diverse perspectives that leverages digital plasticity and configures digital artifacts through ethical and technical decisions.

Still, the value remains for a separable role for user engagement in the development and deployment of digital technology. Even a well-intentioned narrative configuration depends on users to complete the arc of re-signification of actions mediated by technologies: "Without the reader who accompanies it, there is no configuring act at work in the text; and without a reader to appropriate it, there is no world unfolded before the text."[51] These observations by Ricoeur apply also to the configuration of the digital world. We need to return to the rhetoric of the reader and user and explore ways of strengthening users' narrative deliberation. Ricoeur argued in his textual poetics that "another theory of reading is required, one that emphasizes the reader's response—

the reader's response to the stratagems of the implied author."[52] Similarly, in the next chapter, we examine a new form of user relationship with digital technologies that is anchored in the possibilities opened up by developers but whose adoption is informed by the promotion of users' ideals of self-esteem and the good life.

Notes

1 Paul Ricoeur, François Azouvi, and Marc B. de Launay, *Critique and Conviction: Conversations with François Azouvi and Marc de Launay* (New York: Columbia University Press, 1998), 92.
2 Paul Ricoeur, *Oneself as Another* (Chicago: University of Chicago Press, 1994), 269.
3 Ricoeur, *Oneself as Another*, 194.
4 D. Gotterbarn, "Informatics and Professional Responsibility," *Science and Engineering Ethics* 7, no. 2 (April 2001): 221–30.
5 Gotterbarn, "Informatics and Professional Responsibility," 221.
6 Gotterbarn, "Informatics and Professional Responsibility," 227.
7 Paul Ricoeur. "Hermeneutics and the World of the Text," in *Hermeneutics: Writings and Lectures—Volume 2,* translated by David Pellauer (Cambridge: Polity, 2013), 17.
8 David M. Kaplan, "Paul Ricoeur and the Philosophy of Technology," *Journal of French and Francophone Philosophy* 16, no. 1/2 (January 26, 2006): 42–56.
9 Kaplan, "Paul Ricoeur and the Philosophy of Technology," 43.
10 Don Ihde, *Technology and the Lifeworld: From Garden to Earth* (Bloomington: Indiana University Press, 1990), 174-215.
11 Kaplan, "Paul Ricoeur and the Philosophy of Technology," 49.
12 Paul Ricoeur, "The Model of the Text: Meaningful Action Considered as a Text," *New Literary History* 5, no. 1 (1973): 91–117.
13 Ricoeur, "The Model of the Text," 94–95.
14 Without going off on a tangent that would take us away from our argument, it is important to mention that the incongruity between the event of discourse and the written text opens up the field of hermeneutic interpretation and shapes the tripartite relationship between writers, readers, and texts.

15 Ricoeur, "The Model of the Text," 94–95.
16 Ricoeur, "The Model of the Text," 95.
17 Ricoeur emphasizes the link between the duty to remember and the duty to narrate in the face of the horror of Holocaust violence: "This answer is simple and transparent: we must remember because remembering is a moral duty. We owe a debt to the victims. And the least way to pay our debt is to tell and retell what happened in Auschwitz." Paul Ricoeur, *Figuring the Sacred: Religion, Narrative, and Imagination* (Minneapolis: Fortress Press, 1995), 90.
18 Paul Ricoeur, *Time and Narrative, Volume 3* (Chicago: University of Chicago Press, 1988), 157–69 (emphasis added).
19 Wayne C. Booth, *The Rhetoric of Fiction* (Chicago: University of Chicago Press, 1983).
20 Ricoeur, *Time and Narrative, Volume 3*, 249 (emphasis added).
21 Ricoeur, *Time and Narrative, Volume 3*, 166.
22 Ricoeur, *Time and Narrative, Volume 3*, 167.
23 Paul Ricœur, "Architecture and Narrativity," Translated by E. Brennan, R. Carney, and S. Lelièvre. *Études Ricoeuriennes/Ricoeur Studies* 7, no. 2 (2016): 20–30.
24 Alberto Romele, *Digital Hermeneutics: Philosophical Investigations in New Media and Technologies* (New York: Routledge, 2019).
25 Wessel Reijers and Mark Coeckelbergh, *Narrative and Technology Ethics* (New York: Palgrave Macmillan Cham, 2020).
26 Reijers and Coeckelbergh, *Narrative and Technology Ethics*, 85.
27 Reijers and Coeckelbergh, *Narrative and Technology Ethics*, 83.
28 Ricoeur, *Time and Narrative, Volume 3*, 173.
29 Reijers and Coeckelbergh, *Narrative and Technology Ethics*.
30 Ricoeur, *Time and Narrative, Volume 3*, 177.
31 F. Domenico, "Internet, or the Humanistic Machine," in *The Digital Humanist: A Critical Inquiry*, edited by F. Domenico, N. Teresa, and T. Francesca (Santa Barbara: Punctum Books, 2015).
32 Paul Ricoeur, "Ethique et Morale," *Revista Portuguesa de Filosofia* 46, no. 1 (1990): 11. Ricoeur, *Oneself as Another*, 251.
33 Ricoeur, *Oneself as Another*, 273.
34 Ricoeur, "Éthique et Morale," 5–17.
35 Jessica Fjeld, Nele Achten, Hannah Hilligoss, Adam Nagy, and Madhulika Srikumar, "Principled Artificial Intelligence: Mapping Consensus in Ethical and

Rights-Based Approaches to Principles for AI," *Berkman Klein Center Research Publication* no. 2020-1 (January 15, 2020), https://doi.org/10.2139/ssrn.3518482.

36 Europea, Comisión, *Communication from the Commission to the European Parliament, the European Council, the Council, the European Economic and Social Committee and the Committee of the Regions Artificial Intelligence for Europe* (European Commission, 2018), 16.

37 Cédric Villani, Yann Bonnet, Marc Schoenauer, Charly Berthet, Francois Levin, Anne Charlotte Cornut, and Bertrand Rondepierre, *For a Meaningful Artificial Intelligence: Towards a French and European Strategy* (France: Conseil national du numérique, 2018), 114.

38 "Human Rights in the Age of Artificial Intelligence," *Access Now*, 2018, 6, https://www.accessnow.org/wp-content/uploads/2018/11/AI-and-Human-Rights.pdf.

39 Roberto Pereira and Maria Cecília Calani Baranauskas, "A Value-Oriented and Culturally Informed Approach to the Design of Interactive Systems," *International Journal of Human-Computer Studies* 80 (August 1, 2015): 66–82.

40 Pereira and Baranauskas, "A Value-Oriented and Culturally Informed Approach," 66.

41 Ihde, *Technology and the Lifeworld*, 144.

42 Peter-Paul Verbeek, "Materializing Morality: Design Ethics and Technological Mediation," *Science, Technology & Human Values* 31, no. 3 (May 1, 2006): 371.

43 Verbeek, "Materializing Morality," 371.

44 Verbeek,"Materializing Morality," 372–73.

45 Charles Jacobson, Arie Rip, Thomas J. Misa, and Johan Schot, "Managing Technology in Society: The Approach of Constructive Technology Assessment," *Technology and Culture* 38, no. 3 (July 1997): 813.

46 Emanuelle Burton, Judy Goldsmith, and Nicholas Mattei, "How to Teach Computer Ethics Through Science Fiction," *Communications of the ACM* 61, no. 8 (July 2018): 54–64.

47 Emanuelle Burton, Judy Goldsmith, Sven Koenig, Benjamin Kuipers, Nicholas Mattei, and Toby Walsh, "Ethical Considerations in Artificial Intelligence Courses," *AI Magazine*, 2017.

48 Rosalyn W. Berne and Joachim Schummer, "Teaching Societal and Ethical Implications of Nanotechnology to Engineering Students Through Science Fiction," *Bulletin of Science, Technology & Society* 25, no. 6 (December 1, 2005): 459–68.

49 John Wall, "Phronesis as Poetic: Moral Creativity in Contemporary Aristotelianism," *The Review of Metaphysics* 59, no. 2 (2005): 338.

50 Wall, "Phronesis as Poetic," 324.

51 Ricoeur, *Time and Narrative, Volume 3*, 164.

52 Ricoeur, *Time and Narrative, Volume 3*, 167.

5

Users' Sphere of Deliberation

Deliberating Well

In the previous chapter, we discussed the importance of developers employing narrative deliberation to establish a relationship of respect and solicitude toward users who refigure their horizons of meaning using digital technologies. We also recognized that the sphere of developer deliberation must be complemented by the deliberation of users who play an active role in the way they select and integrate digital technologies into their lives. Digital Poetics is a mimetic arc that assumes both the configuration of meaning proposed by developers through digital materiality and the active and deliberate refiguration of meanings by users who interact with digital technologies. As we discussed, user deliberation is essential as a response to the developer's power of configuration; this power of digital technologies can become coercive and engage in rhetorical violence This chapter turns to analyze more directly the process of user narrative deliberation in relation to digital technologies. The concept of "narrative identity" that Ricoeur proposed will structure our investigation of this personal dimension of deliberation. In chapter 3, we presented the ethical deliberation process as an arc that emanates from a retrospective narrative and aims at promoting flourishing lives through a good prospective narrative. We saw that the schema of narrative deliberation can be applied in contexts ranging from an ad hoc decision that parents must make about their children's education to a complex institutional deliberation about the distribution of resources or the evaluation of a judicial sentence. As we will see in this chapter, narrative deliberation can also be productively used to inform decisions on how each person reflects and projects a fulfilled life.

This chapter's central question is, therefore, how to deliberate well from a personal standpoint on using digital technologies. The narrative form, through which Ricoeur considers issues of personal identity, is an excellent guide for our discussion since it allows us to apply the findings of narrative deliberation to each person's life context. In the users' sphere of deliberation, the fundamental object of deliberation is personal identity: how users deliberate about decisions regarding digital technologies affects their life trajectories. Using the language of narrative deliberation, we could reframe this question as to how users can deliberate about their engagement with digital technologies using narrative imagination and incorporate their retrospective narrative identity to foster a flourishing prospective narrative identity. To address this question, we should begin by looking closely into some relevant aspects of Ricoeur's concept of narrative identity, which will raise deeper questions about the impact of digital technologies on users' lives.

Narrative Identity

Ricoeur defines narrative identity as "the sort of identity that human beings acquire through the mediation of the narrative function."[1] This concept emerges as an unexpected offshoot of *Time and Narrative* and becomes one of the cornerstones of Ricoeur's anthropological philosophy, particularly in *Oneself as Another*. I do not intend to offer another exegesis of narrative identity that Ricoeur's commentators have studied profusely in recent decades.[2] Instead, I want to highlight four aspects of the concept that are especially relevant for thinking about the impacts of digital technologies in the sphere of personal deliberation.

First, human identity is a process of constant self-interpretation that finds privileged mediation in narrative form. Narrative identity highlights the mutability of personal identity, which combines forms of temporal permanence with the continuous contextual changes of the self. David Wood writes that the "concept of narrative identity reminds us that our identities are not fixed but continually unfolding."[3] Narrative identity puts in a dialectical relationship, through the concordant discordance typical of the narrative structure, two poles of identity in time that Ricoeur calls idem-identity and ipse-identity. Idem-identity captures the most sedimented traits of our identity, such as our genetic code and ingrained aspects of our character.[4] It is a pole closer to the static identity of other objects in the world. A second form of identity in time, ipse-identity, permeates the identity of the self and captures the eminently human characteristic of selfhood based on one's choices. The paradigm of ipse-identity is the promise through which the self decides to maintain certain identity traits despite the vicissitudes of life. Hence, narrative identity is the product of self-reflection mediated by a narrative approach that attempts to make sense of the spectrum of forms of the self's temporal permanence. Such permanence in time involves more durable traits intertwined with the maintenance of a sense of self in the midst of external and internal change. Such a process is continuous, given that material, environmental, social, and technological conditions continuously alter oneself and one's existential contexts. In *Time and Narrative Volume 3*, Ricoeur says that narrative identity "stems from the endless rectification of a previous narrative by a subsequent one, and from the chain of refigurations that results from this."[5] In our research, we shall focus on how digital technologies impact narrative identities and how they facilitate or obscure the constant task of self-interpretation in the quest for meaning for one's identity.

The second feature of narrative identity we want to highlight is its connection with the ideal of a fulfilled human life. As we have already mentioned, Ricoeur discusses the concept of narrative identity based on Alasdair McIntyre's hierarchy of actions[6] that organizes groups of actions into units of meaning called practices (medicine, chess, farming, painting, music) that are then integrated into life plans (personal, professional, religious) and culminate in a narrative unit of life. Ricoeur finds particular interest in "the connection the narrative makes between estimations applied to actions and the evaluation of persons themselves . . . [N]arrative unity of a life, therefore, serves to assure us that the subject of ethics is none other than the one to whom the narrative assigns a narrative identity."[7] Narrative identity articulates the self that emerges from the narrative interpretation of the ethics of one's actions. Narrative identity provides a basis for evaluating the self's trajectory toward a flourishing life. We will discuss this process of ethical reflection in more detail in the following section, but for now, I highlight the relationship between the hermeneutic unity of narrative identity and its intrinsic connection to the evaluation of a fulfilled life.

For Ricoeur, a fulfilled life is not an individually attainable solipsistic ideal. A life is lived with and for others in just institutions—a famous phrase that encapsulates several aspects of Ricoeurian ethics.[8] Thus, a fulfilled life must have an interpersonal and social dimension. Narrative identity inevitably integrates such dimensions because one's life story can only be told through intertwining one's actions with environmental and institutional events and actions with which one interacts. There is a structural affinity between the multidimensionality (personal, interpersonal, and social) of the ideal of a realized life with the narrative identity that by definition captures such dimensions: "what we call the subject is never given at the beginning. Or, if it is, it is in danger of being reduced to the narcissistic, selfish and stingy ego, from which literature, precisely, can free us."[9]

This structural affinity does not make the effort to accomplish a flourishing life a given. On the contrary, deliberations and decisions are required to asymptotically align the narrated identity with the ideal of a good life that esteems and respects others with whom one interacts face-to-face and with those with whom one shares various institutional spheres, present and future. The subject who narrates her story becomes an agent of her choices and realizes her narrative identity in *praxis*.[10] One interprets one's actions and suffering as a plot worthy of esteem. When later in the chapter we consider the impact of digital technologies on narrative identity, narrative deliberation will be the key to assessing how new capabilities and weaknesses promote or impede the realization of the good life.

Third, narrative identity also references human identity's active and passive poles. On the one hand, narrative identity conveys the notion of a future life plan that projects the stories we want to tell about ourselves. On the other, it integrates the fragility of these projects that external events, unpredictable circumstances, and others' actions affect and shape in ways outside our active choices. Ricoeur summarizes this narrative identity trait through the intertwining of intentions, causes, and chance: "while the notion of life plan places an accent on the voluntary, even willful, side of what Sartre termed the existential project, the notion of narrative unity places its accent on the organization of intention, causes, and chance that we find in all stories."[11] In our context, we need to investigate how digital technologies create new possibilities for intentional actions and also become external causes that impact narrative identity. In both cases, giving new meaning to practices and life plans based on new forms of interactions with digital technologies poses challenges to the self.

Finally, a fourth relevant feature of narrative identity for our discussion is that it is the outcome of a central human capability: "In the reflective form of talking about oneself narratively [*se raconter*] this personal identity is projected as a narrative identity."[12] Narrative identity relies on one of the four fundamental capacities Ricoeur uses to characterize the capable person: speaking, narrating, acting, and imputing the values of actions.[13] Two crucial aspects of the characterization of narrating oneself as a capability require more detail. First, the ability to narrate reminds us that for Ricoeur, humans must be understood primarily as "capable beings."[14] Thus, when digital technologies affect our capabilities, they affect the heart of the Ricoeurian vision of humanity. Second, we need to investigate how digital technologies transform the specific ability to narrate one's narrative identity, since there may be a complex balance between more ability to act and less ability to narrate oneself.

With these fundamental traits of narrative identity as a conceptual backdrop, we have a substantial starting point for our subsequent analysis of the users' sphere of deliberation. Starting from the concept of narrative identity as the overarching theoretical reference, we are ready to approach the impacts of digital technologies in the personal sphere of Digital Poetics.

The Reflexive Arc

In this section, I briefly explore some conceptual building blocks of the "reflexive arc" that structures Ricoeur's *Oneself as Another*, since they provide an organic framework for understanding the connections between narrative identity and the ideal of a good life.[15] The arc offers an essential stepping stone to exploring how to deliberate about digital technologies from the user's perspective. The reflexive arc traces Ricoeur's investigative trajectory, which begins with the analysis of actions and gradually integrates the reflexive movement of the self, which recognizes, projects, and esteems itself based on the meanings it attributes to the actions it takes and suffers throughout its life.

In his dialogue with Strawson in the first chapter of *Oneself as Another* (and in several other works), Ricoeur starts to construct the reflexive arc from the concept of ascription that "marks the reference of all the terms of the conceptual network of action to its pivotal point: 'who?'"[16] Ascription, in the context of action theory, involves a predication of actions or parts of actions to agents.[17] So, for example, the predication "can write an essay" can be ascribed to John or me. Similarly, "can grow tomatoes" can be ascribed to Peter, you, or me.

Ricoeur highlights several limitations with the Anglo-American analytical approach to the concept of ascription, particularly the fact that it refers to any pronominal person and thus cannot capture the distinct experience of the possibility of acting considered from the agent's standpoint. This failure to capture a distinct experience is a desired effect for analytic theory, because it allows the agent to be considered a logical subject like any other and therefore permits the application of the same logical entities to evaluate all attributional operations. Yet this abstraction of the agent's own perspective compromises the reflexive path of narrative identity. "When the self of ascription is omitted, the field is opened for the sameness of the very same thing, to which physical and mental predicates are ascribed. It is this sameness that provides the strength of the argument and that accounts for a portion of the strangeness of our own concept of person."[18] In this logical abstraction, the self is reduced to the same identity as anyone else. This reduction concerns the technology-affected self because, within this limited framework, an analysis of new predications technology enables consideration only of an abstract logical subject (X can communicate via teleconference). Such abstraction cannot address technological affections based on the singular appropriations of meaning of each self that must integrate these new predications into its narrative identity.

To address the agent's primordial position in the conceptual network of action, Ricoeur proposes the concept of attestation as another element of the reflexive arc. Attestation captures the kind of truth proper to the self expressed in the linguistic form "I believe that I can (do something)."[19] The acting self claims the predication of the ascription as its own and is ready to integrate such self-predications into its narrative identity. Thus, a small farmer affirms her selfhood by attesting, "I believe that I can cultivate these vegetables," or a student affirms his selfhood with the attestation, "I believe that I can write this paper."

But attestation is a specific type of truth claim that "characterizes the alethic (or veritative) mode of the style appropriate to the conjunction of analysis and reflection."[20] It arises from the convergence of the analysis of the capacity to act and the reflexive movement of appropriating this act as the result of my power to act. "This means that the search for adequacy between our life ideals and our decisions, themselves vital ones, is not open to the sort of verification expected in the sciences of observation."[21] The veritative mode of attestation requires that the self constantly question its appropriation of actions to reinforce the conviction that it is indeed "capable of." Ricoeur emphasizes that this type of practical assertion ("I believe I can") is stronger than the equivalent theoretical belief assertion ("I believe that . . ."). This difference is linked to the pragmatic evidence that can be listed to support the attestation. Ricoeur speaks of his debt to Jean Nabert's argumentations on this point, which introduces a "detour through the 'what' and the 'how' before returning to the 'who,' which seems . . . explicitly required by the reflexive character of the self."[22]

In later work in *The Course of Recognition*, Ricoeur suggests an intimate relationship between the concepts of attestation and self-recognition and proposes the idea of recognition-attestation that unites the semantic spheres of attestation and recognition.[23] For our particular concern in this chapter, I would like to retain the idea of the self's capacity to attest that certain actions are its own and, in doing so, recognize itself by integrating such capacities to act into its personal identity. Interactions with new technologies generate a new set of questions about the veracity of attestation claims because a new element mediates or interferes with the relationship between agents and their actions. For instance, the statement "I believe I can make new friends" needs to be reconsidered based on the forms of action mediated by social networks. So our guiding question is not only what we can do differently through digital technologies but what do digital technologies do with our personal identity as they interfere in the processes of attestation-recognition. Any suspicion that an attestation is false, that my belief in being able to do something is unfounded, threatens the recognition of self linked to attestation. As mentioned above, Ricoeur highlights the fragility of attestation, which cannot claim the certainty of assertions concerning empirically verifiable things. An intrinsic vulnerability to attestation always exists, which can be amplified or reduced by variations in our ability to act, such as those arising from new digital technologies.[24]

It is also essential to remember that Ricoeur defines attestation as "the assurance of being oneself acting and suffering."[25] This definition reminds us that the capable person is not an all-powerful agent who can always define a priori and execute perfectly all the actions she believes she can perform. For Ricoeur, there is always a pole of fragility in the human constitution that introduces an element of receptivity to the ability to act in the world. In the tenth study of *Oneself as Another*, Ricoeur explores various forms of otherness that are in dialectical tension with personal identity. Among them are the other selves that affect us, hindering, facilitating, preventing, or redirecting our ability to act. Our own body also limits our ability to act and, in some cases, circumscribes our possibilities of attestation. These "others" temper our assurance of acting with fragility. Once again, we have to think about how digital technologies are an "other" that alters the mix between the self's acting and suffering and thus impacts its possibilities of attestation.

In the predicative structure of the action, attestation emphasizes the action's agent who recognizes himself or herself as the subject of the sentence that affirms the possibility of doing X. Besides their relationship to agents, actions may also have ethical qualifications that add attributes such as good or harmful. The attestation process that considers this ethical predication of action allows the agent to recognize himself or herself not only as the efficient source of the action but also as imputable for the moral predications of the action that he or she performs. "Imputability, we shall say, is the ascription of action to its agent, under the condition of ethical and moral predicates which characterize the action as good, just, conforming to duty, done out of duty, and, finally, as being the wisest in the case of conflictual situations."[26] This movement from recognition of being able to act to being ethically attributable by action marks the confluence of two essential capacities of the self.

A second limitation of ascription lies in its temporal neutrality, whose linguistic mark is the modal verb "I can." The fifth and sixth studies of *Oneself as Another* articulate the transition from the logical predications of ascription and reflexive affirmation of attestation to the temporality of the self through the concepts of practices, plans of life, and narrative identity that Ricoeur draws from Alasdair McIntyre's ethics. We have already discussed the concept of practices at various points, but in the reflexive arc that we are traversing, practices determine a more abstract level of attestation that places a series of actions in a teleological relationship ("in view of") recognized as such by a community of practitioners. Thus, someone not only certifies that they can plow the land, but this attestation gains another level of significance when it is considered part of a set of temporally organized activities recognized by a group as farming. Practices have constitutive rules that give meaning to isolated actions based on internal relations to the practice. Ricoeur emphasizes that these constitutive rules add meaning but do not have ethical valence. "[T]hey simply rule over the meaning of particular gestures and . . . make a particular hand gesture 'count as' waving hello, voting, hailing a taxi, and so on."[27]

The concept of "standards of excellence" goes further and captures the properly evaluative and ethical dimension of practices within MacIntyre's conceptual analysis of practices. Based on criteria inherent to the practice, the standards of excellence allow us to evaluate whether a practitioner is a good essayist, a good farmer, or a good chess player. For our concerns about the impact of digital technologies on personal deliberation about a fulfilled life, these standards of excellence matter for two fundamental reasons. First, they establish a criterion of an interpersonal idea of what counts as a good action, which avoids a solipsistic perspective in the ethical evaluation of actions. It is within a community of practitioners that the criteria for recognizing oneself as a good farmer emerge and develop, not on the basis of the isolated observation of an individual. In line with his reflections on the dialectic between tradition and innovation,[28] Ricoeur recognizes that these patterns can change over time. "This cooperative and traditional character of practices does not exclude controversy but instead provokes it, mainly with respect to the definition of standards of excellence, which also have their own history."[29] However, even innovations and challenges to community standards occur based on a baseline of standards of excellence that informs a given practice. Standards of excellence define internal goods immanent to actions within practices and provide an objective—even if always open to possible innovations—anchor to the reflexive process through which those who perform an action recognize themselves as good practitioners based on their actions.

Second, standards of excellence give content to abstract principles of morality. As we will discuss later in this chapter, the reflexive process for Ricoeur culminates in the concepts of self-esteem and self-respect. Self-respect for Ricoeur is self-esteem considered within the formal principles of the moral norm. However, such general principles are devoid of practical content, and standards of excellence provide precisely the "flesh" to the logical structure of deontological principles. What it means for a farmer to act in a way that treats others as ends and never as means—for example, labor policies—should be realized in the form of standards of farming excellence.

Resuming our journey through the reflexive arc that Ricoeur's philosophical anthropology proposes, we reach the level of the "life plan," which encompasses the broader ideals of a fulfilled life, such as professional, family, community, religious, and political life. Ricoeur reminds us that the word "life" in this context must be taken not only in its biological but also in its sociocultural aspects, just as we speak of an intellectual life, a contemplative life, or a business life. The goal is to reach a conceptual level in which the "word 'life' designates the person as a whole, in opposition to fragmented practices."[30] Thus, the actions someone does are organized into practices, and these practices are part of life plans.

At this more general level, the plans take on more abstract contours and extend over long periods, sometimes encompassing an entire life. With this broad view, there is then a game of mutual implication between particular actions, the practices we choose to participate in, and life plans. If up to this point our description of the reflexive arc could be interpreted as a unidirectional and linear path in which we start from more particular, short-term decisions that lead us to more general, long-term decisions, the time has come to break and complicate this analogy. The different structures of the meaning of action are integrated into a constant deliberative process that goes back and forth between basic actions, practices, and life plans. Ricoeur talks about upward and downward movements that are in constant dialectical tension in the sphere of personal deliberation. "[T]he practical field is not constituted from the ground up, starting from the simplest and moving to more elaborate constructions; rather it is formed in accordance with a twofold movement of ascending complexity starting from basic actions and from practices, and of descending specification starting from the vague and mobile horizon of ideals and projects in light of which a human life apprehends itself in its oneness."[31] At the end of this chapter, we shall return to these hierarchical dimensions and consider how they relate to each other in a complex up-and-down movement based on the concept of practical wisdom taken in the sense of "application" that Gadamer suggests. This bidirectional movement will provide insightful connections to narrative deliberation and its application to the digital world.

The notion of a life plan leads us back to our guiding principle of narrative identity. However, our re-encounter with the concept takes place at a different theoretical level, based on the initial segment of the reflexive arc and the concepts of ascription, attestation, practices, and life plans that we have elaborated. This re-engagement with narrative identity in the course of the reflexive arc also points us toward the next segment of the arc in which we will discuss the concepts of self-esteem, self-respect, and a fulfilled life. As we suggested in the chapter's introduction, narrative identity expresses the interpretative process of configuring meaning for one's personal existence based on the intricate network of action units. Narrative identity clarifies that what is at stake is much more than practical projects external to the agent. The entire network of the action's meaning supports the agent's identity and, with it, the perspective of a fulfilled life. Identity, therefore, is not just about what the agent does, how she designs her life plan, and how coherent are her choices about basic practices and actions, for all these nodes or building blocks of the semantic network of action underlie the possibilities of self-recognition and self-esteem.

Within this complex and more nuanced perspective, we can better appreciate how digital technologies impact the different levels of action and constantly challenge the practical calculations that maintain the tenuous and fragile horizon of coherence between life plans, practices, and basic actions. The introduction of mobile devices, for example, affects basic actions such as informal conversation, practices such as medicine, and all dimensions of life involving family, professional, institutional, and political contexts. As a specific example of how digital technologies interfere with the reflexive arc, consider the implementation of software in the playing of board games. In 2016, a computer program using artificial intelligence techniques beat the then-world champion, Lee Sedol, in the game Go,[32] an ancient practice with a vast community of practitioners worldwide. A similar feat had already happened in another practice MacIntyre mentions several times, the game of chess.[33] These examples touch on our questions underlying the trajectory of the reflexive arc that Ricoeur proposes: how do digital technologies create ruptures in the structures of the meaning of action? Before AlphaGo software dethroned Lee Sedol, the same software[34] had already surpassed Fan Hui, the then-European Go champion. One of Fan Hui's reactions encapsulates part of the questions we want to explore in this section. When asked about his defeat, Fan Hui exclaimed in astonishment, "I don't understand myself anymore." By disrupting the practice of the game of Go, digital technology threw into question the practice's standards of excellence and the reflexive movement based on the internal goods immanent in its action. Playing against an agent enabled by artificial intelligence disrupted Hui's attestation, which trickled down to his personal identity. Given this new form of digital technologies with enhanced agency, human agents must deliberate whether and how to integrate this technological disruption into their practices. These examples with board games invite us to investigate the impact of digital technologies in the chain of meaning of other practices, such as in the cases of farmers, doctors, essayists, and so on.

There remains an essential element of the reflexive arc that we must investigate in preparation for the application of user narrative deliberation of Digital Poetics. Our last conceptual station in the reflective arc is self-esteem, which emanates from the complex process of deliberation and self-recognition mediated by actions integrated into practices and life plans that gain ethical significance and are attributed to agents. "Self-esteem draws its initial meaning from the reflexive movement through which the evaluation of certain actions judged to be good is carried back to the author of these actions."[35] If I think that growing vegetables organically is good, and I can do so, I esteem myself as an agent capable of carrying out these actions worthy of positive esteem.

Even if self-esteem originates from the reflexive act in which we value ourselves based on our actions that we consider good, this does not entail for Ricoeur a solipsistic approach to ethics. Self-esteem is too vague and abstract if we do not consider how it relates to our interactions with others. There is an intrinsic dialogic structure of self-esteem. As we shall discuss in chapter 6, even this dialogic structure is incomplete if we do not regard it in the context of fair and just institutions. Our actions, practices, and life plans are realizable only with and for others, whether in face-to-face relationships or sharing public contexts. To recognize and esteem oneself as a producer of organic products, one needs to be in a relationship with other producers, producer associations, agricultural input sellers, and consumers, who could range from family members to third parties buying products. Public service entities might also evaluate one's production. Self-esteem only truly makes sense, then, when we look at it through the lens of three aspects: our actions, interactions with others, and our participation within broader institutions. Furthermore, as the initial anchor of self-esteem is the action evaluated as good, this same type of action is also carried out by other organic producers. Thus, the same transfer process between action and agent must happen in relation to other organic producers who deliberate and act. Therefore, an openness to others is intrinsic to the interpretation of selfhood. Ricoeurian anthropology continually reminds us that "To say self is not to say I,"[36] and that "the passage from selfhood to mineness is marked by the clause 'in each case.'"[37] Self-esteem is distributed among all persons, unlike an egocentric, individualistic form of an "esteem of me." There is a fine dialectic between the distribution of esteem that applies to each person who acts in a certain way and an abstraction that remains latent in the concept of self and that marks the otherness internal to selfhood (ipseity).

A second crucial conceptual nuance needed to properly understand self-esteem within Ricoeurian ethics is that it always has self-respect as its counterpart. While self-esteem marks the teleological dimension of the reflexive movement, self-respect represents the deontological double of this same movement. Ricoeur sees self-respect as the "aspect under which self-esteem appears in the domain of norms."[38] Respect for oneself integrates the idea of norms and laws within the reflective process and, thereby, defies the distortion of self-esteem into a form of "love of myself" that uses others as mere instruments to achieve one's ideal of a good life.

Self-esteem takes us to the end of the reflexive arc. Through this long detour, we have highlighted—certainly not exhaustively—some of the fundamental concepts of Ricoeur's anthropology and ethics for our analysis of the impact of digital technologies on the conceptual network of user action and its ethical consequences. The reflexive arc culminating in self-esteem is based on actions or parts of actions ascribed to agents and experienced as their own through attestation. Narrative identity weaves together actions and agents over time to restitute the temporal dimension and complexity of human ethical experiences. Disruptions in the possibility of these predications propagate architecturally through all the building blocks of the reflexive arc.

What is at stake, therefore, is not only acting differently due to digital technologies but also a change in personal identity and the very ideal of a fulfilled life. A fulfilled life depends on the "capacity of a human agent to designate himself as the author of his acts,"[39] what Ricoeur calls agency. Agency is the basis of self-esteem and self-respect, and these are factors that are structurally related to the self-recognition of one's power to act, to impute to oneself ethical predicates through action, and to project and pursue a flourishing life.

In the next section, we turn our attention to the ways in which digital technologies affect human agency, both expanding the possibilities of acting and complexifying the reflexive movement of self-recognition given the existence of a new form of otherness in collaborative agency that can call into question human agents' attestation and disrupt the reflexive arc.

A New Otherness

In 2023, the Pew Research Center published a report with the suggestive title: "The Future of Human Agency."[40] The report resulted from a survey of more than five hundred technology innovators, developers, business and policy leaders, researchers, academics, and activists who examined how digital technologies empowered by artificial intelligence might impact human agency within a decade or so. The report suggests that "as individuals more deeply embrace these [digital] technologies to augment, improve and streamline their lives, they are continuously invited to outsource more decision-making and personal autonomy to digital tools."[41] In the broad panel of interpretations captured by the report about digital technologies' impact on human agency, some emphasized the increased capacity to act that new technologies bring, and others expressed concern about a possible decrease in human participation in everyday actions and decisions. In general, the panelists seemed to agree that "the current moment is a turning point that will determine a great deal about the authority, autonomy, and agency of humans as the use of digital technology spreads into more aspects of daily life."[42]

Although prediction exercises about the future of technologies are notoriously tricky and permeated by interests and implicit ideological forces, multiple and diversified interpretations of new technologies empowered by artificial intelligence highlight the impact of such technologies on human agency. In this Pew report, the authors barely explore the possible implications of AI's impact on the reflexive arc analyzed in the previous section. Since AI impacts human agency in unique and unprecedented forms, its implications pose a critical question for the development of a deliberative approach to the use of digital technologies in the user's sphere. As we briefly pointed out in chapter 1, artificial intelligence algorithms alter the distribution of agency between humans and nonhumans and therefore affect in a unique way the human reflexive process, whose actions are now primarily performed by a digital "other."

The following paragraphs describe examples of emerging technologies and their potential impact on human capabilities and agency. They analyze the impact on basic biological capacities, and then examine some examples that are closer to the primary question of this chapter, which concerns technologies that directly interfere with the reflexive arc by introducing a new form of otherness that challenges the attestation of the self through reshaping user agency by means of the greater autonomy of digital technologies relative to human agents. The analysis in this section intentionally focuses on forms of "agency augmentation" engendered by shared agency between human and digital agents, leaving to the following sections the analysis of new vulnerabilities in the reflexive arc that accompany these forms of shared agency.

Several newspapers and academic journals have recently reported new advances in brain-computer interfaces. Using biosensors and sophisticated modeling techniques, "[a] completely paralyzed man has been able to communicate entire sentences using a device that records his brain activity. The man was able to train his mind to use the device, which was implanted in his brain, to ask for massages, soup, and beer, and to watch films with his son."[43] A commentator on the work noted, "It's really remarkable to be able to reestablish communication with someone in a completely locked-in state."[44]

Even the ultimate frontier of technological advancements and human embodied fragility, death, is challenged by the new possibilities of digital technologies applied to bioengineering techniques. In 2022, news articles reported, "The Saudi royal family has started a not-for-profit organization . . . that plans to spend up to $1 billion a year . . . supporting basic research on the biology of aging and finding ways to extend the number of years people live in good health, a concept known as 'health span.'"[45] This is hardly an isolated attempt to defy aging and eventually death. In 2013, *Time* magazine asked, "Can Google Solve Death?"[46] reporting on Calico, a Google (technically Alphabet) and AbbVie pharmaceuticals-funded research company. After nine years of existence, Calico states its bold mission and technical capabilities as follows: "Our technology labs work hand-in-hand with scientists throughout the company, advancing hypotheses on how specific technologies and know-how can help answer a research question at the depth required to advance the field. If existing technology does not help us answer a research question, then we'll invent something new."[47] Statements such as this suggest a belief that eventually new technologies will provide the capability to live much longer, if not indefinitely. These growing capabilities open up new possibilities for human agency that will require unprecedented forms of ethical deliberation on the use of such biotechnologies.

Turning our attention to digital technologies that change human agency in everyday life, consider, for example, the mapping applications that have become essential to navigation in urban centers. A collaboration with GPS-enabled applications that perform part of the cognitive process of establishing a route and considering traffic conditions has changed users' ability to go from one point to another. Users can quickly navigate between locations in an unknown city, thus increasing their ability to move around and carry out tasks such as shopping and visiting points of interest.

Recently, another much-discussed example is generative AI-based applications such as ChatGPT for textual media and Dall-E for images. With a few prompts, ChatGPT, Bard, and other text generation applications are capable of creating coherent textual essays with reasonable linguistic and visual consistency. When integrated with other digital tools, such as email applications, this type of technology can read the text of an incoming message and compose a response based on conversational patterns extracted from billions of other messages with which these models are trained. Hence, users can obtain a draft of an email that often speeds up the process of managing their email inboxes. In other cases, it is possible to use these technologies to accelerate content search using a natural language-based interface and receive consolidated information. Image tools such as Dall-E and MidJourney allow sketching images from textual prompts, permitting users with little aptitude for creating digital visual media to express their ideas through digital images.

Messaging and videoconferencing applications enable remote communication, making interpersonal interactions possible at a distance and creating new forms of personal, professional, and academic relationships. During the COVID pandemic, for example, many people maintained their jobs and studies through the new capabilities that emerging digital technologies created. Some of these technologies allow even more sophisticated capabilities, such as recording communication, instantaneous subtitle generation that increases accessibility, and immediate translation that allows inter-linguistic interactions.

Another digital technology growing exponentially in the last decade and enabling new forms of agency is the Internet of Things.[48] It creates capabilities based on the interaction between sensors, mathematical models based on sensor data, and actuators such as mechanical arms or robots that execute commands derived from the mathematical models. These new networks of digital agents allow, for example, the recognition of soil patterns and the optimization of irrigation and fertilizers for a given crop type in a region.[49]

Fostering these and other emergent technologies is a convergence of massive advances in some underlying computational enablers. A brief look at these core enablers gives us a better sense of emergent digital technologies' transformative potential to human agency. The first core enabler we highlight is the increase in processing power that allows us to have in our pockets computers equivalent to NASA's mainframes from a few decades ago.[50] The second core enabler is the internet and its increasing capacity to transmit more significant volumes of data reliably and faster. Enabled by these communication infrastructures, an enormous number of data-gathering devices are constantly capturing data via sensors and human input. For example, as we discussed in chapter 1, mobile devices are continuous sources of data generation that are constantly transmitted over wireless communication networks at high speeds. On the one hand, smartphones are equipped with a myriad of sensors (cameras, GPS, accelerometers, microphones) that automatically collect and transmit data. On the other hand, smartphone users produce a torrent of information as they interact with applications on a daily basis, as is the case with social networks. The gathering and transmission of data would be less valuable if it were not complemented by massive data centers capable of storing and accessing an unfathomable amount of digital information for a relatively low cost of setup and maintenance. Furthermore, a new way of extracting patterns from data collection, called machine learning, makes it possible to create descriptive and prescriptive models about natural and social phenomena, leveraging the digital memory stored in databases globally. Finally, mechanical advances and bioengineering allow the creation of actuators that modify material and biological structures from the models that machine learning algorithms create.

The convergence of processing, sensing, communication, mnemonic, algorithmic, and acting enablers creates a synergetic effect that explains the unique new capabilities digital artifacts mediate and prompts us to further inquire into the impact of this new digital agency balance on the human reflexive arc.

Digitally Mediated Agency

With a focus on various examples of technologies that embody these new digital enablers and have unique ways of interacting in the world with increasing autonomy, this section introduces concepts and arguments from philosophers of technology who have discussed new forms of mediation of action that the technologies provide. These brief descriptions focus on the emergence of new forms of agency through the collaboration between humans and technologies, tracing a spectrum of the diverse manifestations of this intricate mix of agency which requires constant reinterpretation of who the agent is. While some of the examples depart from our emphasis on digital technology, they properly illuminate the effects of digital technology on human agency. The final part of this section suggests how Ricoeur's reflexive arc allows us to coherently extend these discussions from the pragmatic plane to the hermeneutic and ethical plane. Let's start this inquiry with Bruno Latour's pragmatic approach.

Latour suggested the analysis of action based on "actants." Actants are co-agents in technologically mediated actions. Actants instead of agents allow Latour to focus on the hybrid nature of technologically mediated actions that result from a joint agency of humans and nonhumans. For example, discussing the various interpretations of agency in the use of automatic firearms, Latour claims that "we must learn to attribute—redistribute—actions to many more agents than is acceptable to either the materialist or the sociological account."[51] For him, both the materialist view that firearms are determinants of action and the sociological view that all agency remains with the user of technology start from the wrong principle of thinking of actants as isolated essences. They ignore that in the new interaction between humans and technologies, a novel form of agency emerges, and, more relevant to our discussion, the actants are transformed. "You are a different person with the gun in your hand."[52]

Another interesting example for our purposes that Latour offers is the difference in agency between laboratory technicians and precision chemistry practitioners when using different types of pipettes. Latour juxtaposes the historic pipette Pasteur used a century ago and the contemporary, automated pipette called the "Pipetman." The first requires meticulous measurement. This measurement process involves a scrupulous examination of the clear glass and aligning the liquid level with the tiny precision measurements engraved on the glass surface. Each Pasteur pipette immersion in the fluid before transferring it to a different container requires extraordinary care by laboratory technicians. In contrast, to operate the modern Pipetman pipette, the technician simply presses the top of the pipette twice—once to suck in the liquid and once to expel it—and manipulates the buttons on the top of the instrument to adjust the volume withdrawn with each immersion. Although both pipettes require that technicians have a set of skills and actions, the distribution and nature of these skills differ significantly. The Pasteur pipette requires technicians to exercise an exceptional degree of coordination and mastery for each new immersion. With the contemporary pipette, the technology largely carries out the measurements multiple times based on the device's settings. Latour describes the impact of this transition: "the total sum of activity—comparing my relation to the Pasteur pipette with my relation to the pipetting robot—is maintained or increased, but its distribution has been modified. Some highly trained technicians are made redundant, unskilled workers are recruited, high-tech companies are created in order to produce robots where simple workshops were until recently sufficient."[53] In this pipette transition, the technician's agency changes substantially, and, considering the reflexive arc, we must ask ourselves about the implications of such a shift for the chemist's or laboratory technician's self-recognition and self-esteem. It is also clear that the practice's standard of excellence needs to adapt as this specific action of precision transfer of liquids is no longer indicative of the technician's excellence. In its place emerges a new capability related to the configuration, adjustment, and maintenance of sometimes rather complex features of the robot.

Certainly, researchers still need to know how to configure, program, and use the robot, but we want to emphasize that the increasing digital agency leads to a shift in the human reflexive arc due to this change in the distribution of basic research laboratory actions. It is reasonable to think that with advances in artificial intelligence, additional parts of the pipetting process will be transferred to technological actants. For instance, an AI-enabled robot can learn the most effective quantity of liquid for a given laboratory test and adjust itself to optimize the procedure, diminishing the need for additional human intervention in the setting and adjustment sub-processes. This example illustrates how the greater autonomy of digital technologies creates a new cycle of disruptions in collaborations between humans and nonhumans. Furthermore, the new set of research activities related to the design and implementation of the robot itself creates new human activities and establishes new practices. Such activities constantly create new challenges for narrative deliberation in the users' sphere, such as educational investment choices, professional career possibilities, and ultimately what activities can foster self-esteem and a fulfilled life.

Latour's study on human agency corroborates two fundamental points for our analysis of narrative deliberation in the personal sphere. First, interactions with technologies create a new spectrum of agency. New technologies, particularly those, such as AI, that increase technological agents' autonomy through the ability to learn patterns and perform predictive tasks, dramatically impact the mix of agency between humans and digital technologies. Thus, the concepts of ascription and attestation that support the reflexive movement of self-esteem need to be constantly reinterpreted according to new forms of digital collaboration.

Second, Latour's attention to the particular and malleable form of human identity allows an approximation with Ricoeur's notion of narrative identity and the pole of ipse-identity that the changing circumstances of the self constantly challenge. Latour's emphasis on the impact of technologies on the circumstances that affect the participation of the human agent is fascinating for our argument. Returning to the example of the pipette, Latour describes, for example, the process of distancing in which the technicians no longer recognize themselves as authors of laboratory measurements: "Once I have bought the calibrated Pasteur pipette, I can then go on with my skilled task. Once I have turned the knobs of the automatic pipette, I can then fall back on a less skilled task. The enunciator, in other words, may be absent itself. Even my own action of a moment ago is now foreign to me, although still present in a new guise."[54] This rupture between the enunciator and the agent indicates the impact of digital technologies on the narrative identity of the agents who no longer recognize themselves as the author of this or that action. There is an estrangement between the self-narrator and the agent of the action.

The redistribution of agency affects the human agent's ability to narrate itself from the measurement activity that, before the introduction of the automatic instrument, was an integral part of the laboratory technician's narrative identity.

Don Ihde is another philosopher of technology who has extensively explored how technologies are intrinsically linked to human actions and impact the analysis of how we act and recognize our actions. Ihde, who studied and published on Ricoeur's work,[55] developed his arguments in several books and articles within the field that became known as post-phenomenology. In one of his best-known works, *Technology and the Lifeworld: From Garden to Earth*,[56] Ihde proposed a fruitful taxonomy of interactions between humans and technologies organized into four main categories of relationships.[57]

The first form of interaction between humans and technology Ihde identified is a mediation relationship, where a technological artifact influences our perception. This means that, instead of perceiving the world directly, we perceive it through technological means, similar to when we wear glasses or watch television. Ihde divides mediation interactions into two subtypes. The first is described as incorporation relations. In this type of interaction, individuals incorporate technological artifacts into their experience, effectively expanding their bodies' sensitivity to the world. The use of glasses serves as an example of this relationship of incorporation because when someone wears glasses, the focus is not on the glasses themselves but on viewing the world through them.[58] Such technological artifacts become almost transparent mediators of the way users perceive the world. In the digital sphere, we could think of earbuds, which allow us to hear sounds transmitted remotely and create a new interaction between our hearing and the world. Ihde proposed a second subtype of mediation interaction with technological artifacts that he calls hermeneutic relations.[59] In these relationships, we engage with the world through an artifact, but this artifact is not transparent, as it was in the prior example of embodiment. Rather than withdrawing from our connection to the world, the artifact offers a representation of the world that requires interpretation to convey some understanding. An example Ihde gives is the thermometer, which shapes a new relationship with the world through a model that needs user interpretation.[60] In the case of digital technologies, smartwatches are an excellent example of new technology that mediates our relationships with the world and ourselves but need to be interpreted so that this new perception of the world makes sense to the user.

These examples of mediation technologies and the concepts Ihde proposed are particularly fruitful for our discussion on human agency because they highlight how technologies affect our actions, including the ways we perceive the world, which in turn alters how we act upon our perceptions. Thus, doctors share agency with all the imaging equipment—such as ultrasounds and endoscopes—that they use in their diagnostic process, and imaging equipment mediates their perception of biological and physiological processes. Digital technologies such as high-resolution magnetic resolution imaging amplify and transform the agency of doctors who can "see" morphological and functional aspects of the body and thus alter their *actions* and *decisions* regarding the course of treatment or possible surgical interventions.[61]

The second form of interaction Ihde proposed establishes a relationship of alterity, which is not a relationship to the world through an artifact but rather a relationship with the artifact itself.[62] In relationships of alterity, the emphasis shifts from a relationship with technologies in order to experience the world to the relationship with the technological artifact that renders the relationship with the world opaque. Of much interest to our discussion in this chapter is Ihde's claim that the role of technology in these relationships can be described as a "quasi-other." Returning to our image diagnosis example, now doctors can interact using natural language with technological artifacts that analyze images and indicate potential health problems. Digital artifacts thus become quasi-others with which doctors interact, transforming their agency and transferring part of their actions to the technological agent. However, this relationship of otherness impacts our analysis of agency even more when technological agents interact directly with users/patients. With the increasingly frequent use of artificial intelligence to recognize patterns in images, activities once mediated by a human doctor are carried out by autonomous digital agents. Applications on mobile devices that capture skin images can indicate with a high degree of precision the pathology risk present in the image.

Finally, Ihde refers to a third type of relationship that he calls background relationships, where technological artifacts shape our perception of reality while remaining imperceptible, much like thermostats that adjust temperature without requiring our intervention or awareness.[63] These technologies discreetly switch between active and passive states in the background of our daily lives, going almost unnoticed. They create an ambient "field" that allows us to interact and have experiences without us ostensibly being aware of their existence. They have a simultaneous presence and absence: without our conscious recognition, they shape our experiences, providing a structure for them.[64]

These background relationships of technologies gain more agency with advances in coordination among digital artifacts that can perform sophisticated processes of collaboration. A more recent example is Smart Homes, in which not only do thermostats act in the background, altering our field of perception, but devices talk to each other, creating a network of the Internet of Things that makes decisions and acts by transforming our experiences. Thus, it is possible to create a wake-up routine in which a stereo device plays one of our favorite songs and then notifies us of our appointments, the coffee machine prepares our first dose of daily caffeine while the room temperature is controlled for our morning comfort, and "smart" curtains adjust based on external lighting sensors. Again, this form of background relationship points us to a series of everyday actions once performed exclusively by human agents and now, through digital technologies, are "outsourced" to these autonomous agents. The peculiarity of this relationship is that we do not even recognize our interaction with the technologies that present themselves as aspects of the world of life.

This brief commentary on Don Ihde's phenomenological analysis of technological mediations has pointed to several ways in which digital technologies impact the reflexive arc we discussed in the previous section, from how we perceive the world, how we act through and with digital others and even how we are replaced by digital agents.

Ernst Wolff has also studied the various forms in which human action is permeated by its collaboration with other technical things. In his book, *Between Daily Routine and Violent Protest*, he develops a phenomenological-hermeneutic perspective in dialogue with Ricoeur that deeply explores the technicity of action.[65] Wolf defines technicity as the "combination of capabilities and means" and discusses how ethical expectations and hesitations permeate these intrinsic dimensions of human action. He argues that "the capable human *cannot* be a capable human *without* being a technical human."[66] Although Wolff broadens his concept of technicity to methods of acting, his fundamental thesis highlights the compositional aspect of human agency: "Different individual, organizational or institutional agents combine actional capabilities with tools, institutions, infrastructure, etc. by means of which they act."[67]

All these thinkers in the philosophy of technology point to the practical impacts of technologies on human agency and have highlighted some possible ethical repercussions of such changes in the scope of human action. All these impacts on human agency require ethical deliberation that takes into account their impacts on narrative identities and self-esteem.

In the next section, we will see how an approach from Ricoeur's hermeneutic philosophy adds to these pragmatic and epistemological approaches to form an organic view of the ontological effects of digital technologies based on the conceptual network that arises from the idea of the capable person. By placing the impacts of agency digital technologies in the context of Ricoeurian philosophical anthropology and the reflexive arc outlined above, the new requirements for ethical deliberation in the users' sphere become evident and further emphasize the importance of narrative deliberation for technology issues.

Aiming at a Good Life

Disrupting Attestation

Ricoeur insists on the fragility of attestation as "attestation of selfhood is inseparable from an exercise of suspicion." He lists some of the forms in which attestation may fail, such as by "lies, deceit, misunderstandings, and illusions."[68] As suggested at the beginning of this chapter, emergent digital technologies introduce new forms of otherness within the conceptual network of action, complexifying the predicative structure of ascription. These collaborative ensembles of humans and machines create a new dimension of suspicion that might fracture the reflexive arc, making the self question its attestation since now digital (co)-agents take over so much of the action (or so many actions of one's *praxis*). Can one really believe one can farm since the Internet of Things automated devices do most of the seeding, watering, pest control, soil tending, and harvesting? Can one really believe one can write an essay since ChatGPT or Bard do most of the word selection, content research, and paragraph organization? Can one really believe in one's capability to make and maintain friendships if almost all of these personal relationships are mediated by algorithms, notifications, and social media application interfaces?

Ricoeur's philosophical anthropology presented in *Oneself as Another* prompts us to investigate further the ethical analysis of the capabilities that digital technologies engender. This inquiry goes beyond most of the current philosophy of technologies, which focuses primarily on pragmatic and epistemic matters. As in the internal dynamics of *Oneself as Another*, we need to rescue attention from the limited focus on the duality of the questions "why?" and "how?" and to direct ourselves to the deeper anthropological and ontological question of "who?"

Digital technologies deeply impact the reflexive arc, not only disrupting what people do, how they do it, and how they understand the world but also how they understand and value themselves and give meaning to their existence. The new level of agency that artificial intelligence-based technologies grant raises ontological and ethical questions about the possibility of the self narrating and recognizing itself through actions that, until recently, were exclusively attributed to human agents. Digital technologies' possible rupture in attestation propagates its effect on narrative identity and self-esteem and compromises the ideal of living a good life.

Traversing the Reflexive Arc Backward through Narrative Deliberation

The analysis of the impact of digital technologies on end users' agency, based on Ricoeur's philosophical anthropology, advises us to retrace the reflexive arc backward so that the ideal of a good life anchored in narrative identity and self-esteem becomes the foundation—and not the victim—of any attempt to adopt and use digitally mediated technologies that might take away from ourselves opportunities to act meaningfully. This approach also fosters the intentional integration of digital technology so that it can become an agentive amplifier, carefully aligned from the very beginning with the human agent's perspective of a flourishing existence.

In this sense, the narrative deliberation that we proposed in the third chapter allows the self to make decisions about when and how to integrate digital technologies, and which ones, into its narrative identity based on the ideal of a good life with and for others in just institutions. Narrative deliberation interrupts the automatism of thoughtless adoption of new technologies and invites users to consider the ways in which these new artificial agents impact one's prospective narratives. This consideration does not mean a total denial of the possible benefits of integrating new technologies but a teleological approach that starts from estimating the impacts of new technologies on a prospective narrative identity.

As we have seen throughout this chapter, digital technologies increasingly are gaining space in the balance of human-machine collaboration. The current trend, driven by the market's impulse to increase the number of users, has been one of uncritical assimilation of technologies into users' lives. Narrative deliberation demands and promotes a pause in the "automatic automation" of human actions with the sole intention of more execution efficiency. By deliberating on the possible impacts of technologies on narrative identity, users reclaim the fundamental agency of the process by aligning technological possibilities with their ideals of a good life rather than seeking ways to have fulfilled lives *despite* the impacts of digital technologies. It is in this sense that we propose that the reflexive arc should be traversed from back to front, always starting with the ideas of narrative identity and self-esteem, to gradually deliberate on how digital technologies should be integrated into practices and particular actions to ensure the desired impacts on the attestation and ascription of actions shared with technological agents.

Students, for example, can deliberate how, when, and for which tasks they use large language models (LLMs) in their learning process. Starting from their ideal of a fulfilled life and considering the possible impacts of using LLMs on their prospective narrative identity and the self-esteem they derive from the reflexive process of their academic work, students can apply narrative deliberation to decide how to integrate the agency of computational entities such as ChatGPT and Bard in their learning actions. Through narrative deliberation, students can analyze retrospective narratives about which strategies they used to learn fruitfully and how and whether new forms of digital otherness can promote a prospective narrative identity aligned with their goals of a fulfilled life.

In turn, small agricultural producers should deliberate on which and for what purposes digital technologies should be integrated into their practice, starting from the ideal of a fulfilled life, evaluating the changes that certain technologies will bring to their agency, and how this could eventually impact their self-esteem and the balance of ascription changes. For this narrative deliberation, agricultural producers must analyze their prospective narrative, possibly using imaginative variations based on similar narratives of technology application in other real or fictional contexts. The notion of phronetic plurality intrinsic to narrative deliberation is also fundamental in this scenario so that the possibilities and technical risks certain artifacts bring are properly considered in the deliberative process of the small agricultural producer who decides, for example, on the potential adoption of an automated irrigation or pest control system.

Social media users can also react to the almost irresistible attraction of their feeds[69] through narrative deliberation that traverses the reflexive arc backward. Instead of simply having their agency hijacked by notifications of new posts and recommender systems that capture attention with increasing precision, users can deliberate on the most appropriate ways of integrating social networks based on the ideal of a good life. For example, suppose the impulse to receive new information from social media affects the agency of being fully present in conversations with friends and family. In that case, users may deliberate about adjustments to their interaction with digital artifacts so that they can again recognize and derive self-esteem from being a friend and family member attentive to the questions and concerns of people physically close to them.[70]

In all these examples, Ricoeur's narrative theory allows users to consider the possible ontological impacts of the technological mediations of human action as described by Latour, Ihde, and Wolff and to integrate them into a process of ethical deliberation. Imaginative variations that explore how technologies have affected historical or fictional narratives alert users to possible unwanted breaks in the reflexive arc. Narrative imagination enables users to weigh up what roles they want to assign to digital technologies in the fabric of their narrative identities. And the critical practical wisdom intrinsic to narrative deliberation enables users to apply their moral principles in new situations brought about by the new possibilities of digital technologies. The deliberative process anchored in the concept of narrative identity puts the user back at the teleological center from which technological mediations and digital others can be intentionally chosen as productive and effective means. Narrative deliberation in the users' sphere ensures that integrating new technologies starts from the ideal of a fulfilled life and goes backward, analyzing how and whether digital technologies should be integrated into users' actions.

Notes

1 Paul Ricoeur, "Narrative Identity," *Philosophy Today* 35, no. 1 (1991): 73.

2 See, for instance, David Wood, *On Paul Ricoeur: Narrative and Interpretation* (London and New York: Routledge, 2002). Alberto Romele, "Narrative Identity and Social Networking Sites," *Études Ricoeuriennes* 4, no. 2 (January 3, 2014): 108–22. Annemie Halsema, "The Time of the Self: A Feminist Reflection on Ricoeur's Notion of Narrative Identity," *Time in Feminist Phenomenology*, edited by Christina Schües, Dorothea Olkowski, and Helen Fielding (Bloomington: Indiana University Press, 2011): 111–31.

3 Wood, *On Paul Ricoeur*, 11.

4 Paul Ricoeur, "Life in Quest of Narrative," in *On Paul Ricoeur. Narrative and Interpretation*, edited by David Wood (London and New York: Routledge, 1991), 21.

5 Paul Ricoeur, *Time and Narrative, Volume 3* (Chicago: University of Chicago Press, 1988), 248.

6 Alasdair MacIntyre, *After Virtue* (Notre Dame: University of Notre Dame Press, 1981).

7 Paul Ricoeur, *Oneself as Another* (Chicago: University of Chicago Press, 1994), 178.

8 Ricoeur, *Oneself as Another*, 172.

9 Ricoeur, "Life in Quest of Narrative," 33.
10 Ricoeur, *Time and Narrative, Volume 3*, 149.
11 Ricoeur, *Oneself as Another*, 178.
12 Paul Ricoeur, *The Course of Recognition* (Cambridge, MA: Harvard University Press, 2007), 99.
13 Ricoeur, *Oneself as Another*, 16.
14 Ricoeur, *Oneself as Another*, 181.
15 In her insightful keynote presentation at the Fonds Ricoeur's 2023 Summer Workshop, Eileen Brennan referred to the "reflexive arc" that structures Ricoeur's *Oneself as Another.* E. Brennan, "Parting Ways with the Philosophies of the Subject: Radical Break or Nothing New?" in *Sixth Edition of the Fonds Ricoeur's Summer Workshop* (Paris: Fonds Ricoeur, 2023).
16 Ricoeur, *Oneself as Another*, 27.
17 Ricoeur, *Oneself as Another*, 35.
18 Ricoeur, *Oneself as Another*, 36.
19 Ricoeur, *Oneself as Another*, 21.
20 Ricoeur, *Oneself as Another*, 21.
21 Ricoeur, *Oneself as Another*, 180.
22 Ricoeur, *The Course of Recognition*, 93.
23 Ricoeur, *The Course of Recognition*, 91–92.
24 Ricoeur, *Oneself as Another*, 22.
25 Ricoeur, *Oneself as Another*, 22.
26 Ricoeur, *Oneself as Another*, 292.
27 Ricoeur, *Oneself as Another*, 155.
28 We will explore this dialectic further in the institutional sphere of Digital Poetics.
29 Ricoeur, *Oneself as Another*, 176.
30 Ricoeur, *Oneself as Another*, 177.
31 Ricoeur, *Oneself as Another*, 158.
32 Elizabeth Gibney, "What Google's Winning Go Algorithm Will Do Next," *Nature* 531, no. 7594 (March 17, 2016): 284–85.
33 Andrew McAfee and Erik Brynjolfsson, "Opinion," *New York Times*, March 16, 2016, https://www.nytimes.com/2016/03/16/opinion/where-computers-defeat-humans-and-where-the y-cant.html.
34 Brenden Lake, "Finding Solace in Defeat by Artificial Intelligence," *MIT Technology Review*, April 28, 2017, https://www.technologyreview.com/2017/04/28/106009/finding-solace-in-defeat-by-artificial-intelligence/.

35 Ricoeur, *Oneself as Another*, 172.
36 Ricoeur, *Oneself as Another*, 180.
37 Ricoeur, *Oneself as Another*, 180.
38 Ricoeur, *Oneself as Another*, 171.
39 Paul Ricoeur, *The Just* (Chicago: University of Chicago Press, 2000), 2.
40 Janna Anderson and Lee Rainie, *The Future of Human Agency* (United States of America: Pew Research Center: Internet, Science & Tech, 2023), https://policycommons.net/artifacts/3456462/the-future-of-human-agency/4256835/.
41 Anderson and Rainie, *The Future of Human Agency*, 4.
42 Anderson and Rainie, *The Future of Human Agency*, 5.
43 Jessica Hamzelou, "A Locked-in Man Has Been Able to Communicate in Sentences by Thought Alone," *MIT Technology Review*, March 22, 2022.
44 Hamzelou, "A Locked-in Man," 2022 (quoting Jaimie Henderson, a Stanford University neurosurgeon).
45 Antonio Regalado, "Saudi Arabia Plans to Spend $1 Billion a Year Discovering Treatments to Slow Aging," *MIT Technology Review*, June 7, 2022. https://www.technologyreview.com/2022/06/07/1053132/saudi-arabia-slow-aging-metformin/.
46 Harry McCracken and Lev Grossman, "Google vs. Death," *Time*, September 30, 2013, https://content.time.com/time/magazine/article/0,9171,2152422,00.html.
47 "Research & Technology," *Calico*, Accessed October 30, 2023, https://calicolabs.com/research-technology.
48 See, for instance, Samuel Greengard, *The Internet of Things, Revised and Updated Edition* (Cambridge, MA: MIT Press, 2021).
49 Shubo Liu, Liqing Guo, Heather Webb, Xiao Ya, and Xiao Chang, "Internet of Things Monitoring System of Modern Eco-Agriculture Based on Cloud Computing," *IEEE Access* 7 (2019): 37050–58.
50 Neil C. Thompson, Shuning Ge, and Gabriel F. Manso, "The Importance of (Exponentially More) Computing Power," *arXiv [cs.AR]*, June 28, 2022.
51 Bruno Latour, "On Technical Mediation," *Common Knowledge* 3, no. 2 (1994): 32.
52 Latour, "On Technical Mediation," 32–33.
53 Latour, "On Technical Mediation," 33.
54 Latour, "On Technical Mediation," 45.
55 See, for instance, Don Ihde, "Expanding Hermeneutics," in *Hermeneutics and Science: Proceedings of the First Conference of the International Society for Hermeneutics and Science*, edited by Márta Fehér, Olga Kiss, and László

Ropolyi (Dordrecht: Springer Netherlands, 1999). Don Ihde, *Hermeneutic Phenomenology: The Philosophy of Paul Ricoeur* (Evanston: Northwestern University Press, 1971).

56 Don Ihde, *Technology and the Lifeworld: From Garden to Earth* (Bloomington: Indiana University Press, 1990).

57 Although Ihde's examples are not necessarily drawn from digital technology, they provide, nonetheless, relevant conceptual insights for our investigation of Digital Poetics.

58 Ihde, *Technology and the Lifeworld*, 47.

59 Ihde, *Technology and the Lifeworld*, 63.

60 Ihde, *Technology and the Lifeworld*, 84–85.

61 See, for instance, the application of magnetic resonance imaging to investigate intracranial arterial diseases: Bodle et al., "High-Resolution Magnetic Resonance Imaging: An Emerging Tool for Evaluating Intracranial Arterial Disease," *Stroke* (2012), https://doi.org/10.1161/STROKEAHA.112.664680.

62 Ihde, *Technology and the Lifeworld*, 97.

63 Ihde, *Technology and the Lifeworld*, 108.

64 Peter-Paul Verbeek, "Don Ihde: The Technological Lifeworld," in *American Philosophy of Technology: The Empirical Turn* (Indiana University Press, 2001), 131–32.

65 Ernst Wolff, *Between Daily Routine and Violent Protest: Interpreting the Technicity of Action* (Berlin-Boston: De Gruyter, 2021).

66 Wolff, *Between Daily Routine and Violent Protest*, 110.

67 Wolff, *Between Daily Routine and Violent Protest, 3.*

68 Ricoeur, *Oneself as Another*, 72.

69 Adam Alter, *Irresistible: The Rise of Addictive Technology and the Business of Keeping Us Hooked* (New York: Penguin, 2017).

70 In this chapter, I focus on the possibilities of deliberate user agency, but as the example of social networks demonstrates, it is necessary to take into account emotional aspects that are not fully linked to rational deliberation, particularly in fragile or still developing user groups. This essential aspect of the impact of digital technologies on human agency points to the need for an institutional sphere of deliberation that complements the deliberative sphere of users. We will consider the institutional sphere of deliberation in the next chapter.

6

The Institutional Sphere of Deliberation

Institutions

In the previous chapters, we discussed how developers and users should apply narrative deliberation when designing and applying digital technologies. Developers motivated by a relationship of respect and solicitude for users should deliberate on designing and implementing new artifacts that configure a prospective narrative that shapes a digital world and promotes conditions for users' fulfilled lives. Users, in turn, must take on an active deliberative role and decide, based on their ideals of a fulfilled life, self-esteem, and self-respect, how (and if) to refigure their narrative identities based on the digital worlds proposed by the developers of new technologies. However, the pervasiveness of the digital world, the increasing complexity and agency of digital artifacts, and the economic structures that mediate the relations of development and distribution of new technologies demand deliberation and agency from institutions to ensure that the digital world promotes ethical ways of refiguring experiences and meanings for users, particularly those from fragile groups whose capacity for deliberation can be limited in the face of the intrinsic complexities of digital technologies. This chapter discusses the ethical challenges and application of narrative deliberation in the institutional sphere of Digital Poetics.

The concept of "institutions" is polysemic and used with different objectives and semantic contours in different contexts. Hence, we must start by delineating how we use it in this chapter. Ricoeur defines the concept of institution as "the structure of living together as this belongs to a historical community—people, nation, region, and so forth,"[1] and he qualifies it by saying that "what fundamentally characterizes the idea of institution is the bond of common mores and not that of constraining rules."[2] Thus, the meaning of institution is broader than just a type of political or legislative body, and the essential attribute of institutions is "living together" on the basis of "common mores." Therefore, narrative deliberation at the institutional level must consider the ways in which digital technologies impact the common mores of communities pursuing a fulfilled life shared with others.

The scope of the concept of "institution" significantly broadens and complexifies the ethical issues of this sphere of deliberation, as it is not restricted to a well-defined context of collective deliberation. Within this sphere, we must consider, for example, how associations of small farmers will adopt digital technologies to optimize harvesting and pest control; how schools and universities will integrate AI-based assistants into pedagogical processes; how companies will integrate digital technologies into their processes for selecting and retaining professionals; and certainly how legislators and judges will create and apply new laws to regulate emerging digital technologies.

In recent years, there has been a growing movement by national and regional institutions to create public policies and regulate legally emerging digital technologies such as social media, short video applications, and artificial intelligence, which makes this context a reasonable starting point for discussions in the institutional sphere.[3] Several proposals and initiatives in this legislative context focus on establishing rules, guidelines, and ethical principles to guide the responsible development and deployment of new technologies in order to minimize possible negative consequences. Some of the most common themes of current efforts at institutional deliberation on digital technologies include data privacy, algorithmic biases, the spread of disinformation, technological dependence, impacts on jobs, excessive concentration of power and wealth, and long-term existential risks.[4] Among the proposals to address the ethical problems arising from emerging technologies are government regulations for supervising and auditing AI systems, data protection and privacy laws, content moderation policies for platforms, labeling of AI-generated content, and other fundamental principles designed to avoid the harmful effects of digital systems.

Yet, narrative deliberation should go beyond establishing general principles and abstract guidelines. As we have argued repeatedly, narrative deliberation should integrate legal and ethical principles into a broader ethical decision-making arc that is constantly guided by the development of a forward-looking narrative that maximizes the possibility of community members having fulfilled lives. There are two salient benefits of applying narrative deliberation in the institutional sphere. First, narrative deliberation provides a theoretical framework in which a plural debate explores the consequences of new digital technologies in the various social contexts in which they will be applied. Second, narrative deliberation invites a positive approach to the ethical impacts of emerging technologies and not just a negative discussion of what should not be done, although, as we noted in chapter 3, this is a constitutive part of narrative deliberation.

Thus, discussion and deliberation in the institutional sphere should encompass, but not be limited to, preventing technologies from promoting misinformation, addictions, mental illness, and polarization. In the context of narrative deliberation, the fundamental question is how digital technologies promote the good life in just institutions while avoiding the pathologies that can be byproducts or undesired effects of technologies. This shift toward attention to the good life offers a radical change in focus and approach since the starting point of the discussion is the prospective narrative that the new technology affects and how it may or may not increase the quality of life of the people impacted by such technologies.

In each sphere of deliberation, we adopt structuring conceptual components that are fundamental to the teleological vision of the good life, which guides narrative deliberation. These components vary according to each group involved but have in common a direct relationship with Ricoeur's philosophical anthropology. For users, we emphasized in chapter 5 narrative identity, self-recognition, and self-esteem through the reflexive arc. As for developers, in chapter 4, we highlighted respect and solicitude toward users, for whom Digital Poetics creates a digital world in which users may refigure their lifeworlds. In this chapter, we stress the promotion of mutual recognition as a guideline for positive decisions related to the regulation, investment, and adoption of digital technologies in the institutional sphere.

Therefore, the guiding question for this chapter is: How should institutions deliberate in order to promote capacity building and mutual recognition through the introduction of new technologies?

From Self-Recognition to Mutual Recognition

In the previous chapter, we focused on the application of narrative deliberation by users of digital technologies based on the concept of self-esteem. We proposed that the reflexive arc that leads from the ability to act to the ideal of a fulfilled life through self-esteem must be traveled backward. Narrative deliberation must start from the ideal of a good life and self-esteem in order to choose how, when, and toward which goals users should integrate digital technologies into their lives. In this chapter, we want to unfold the goal of a good life in the public sphere through the concept of mutual recognition. Thus, if in the users' sphere, digital technologies should be considered from the point of view of how they impact personal narrative identity and *self*-recognition, in the institutional sphere, deliberations on digital technologies should be guided by the promotion of *mutual* recognition. Based on a collective prospective narrative aimed at mutual recognition, decisions about which technologies to promote and where to apply them become intentional and increase the possibility of communities developing capacities equitably through novel technological artifacts. This framework proposes a sharp departure from a certain form of techno-solutionism[5] that appears in some public policies or from the quasi-imposition of social transformations based exclusively on the economic interests of large technology companies.

Ricoeur suggested this transitional conceptual path between the individual and the institutional through the concept of recognition in the final part of his book, *The Course of Recognition.*[6] We will return to this text but want to set the stage by paying attention to his essay on "Capabilities and Rights" from the same time period. There Ricoeur intended to investigate an epistemological and existential foundation that could connect the concepts of capability and rights, the former related to philosophical anthropology and the latter to a philosophy of law. He proposed the idea of recognition as a subliminal link between capabilities and rights and, therefore, between the personal and public spheres. He developed this conceptual network by taking recognition as "a dynamic process connecting a plurality of points of view as the distinctive steps of the same development."[7]

In the first step of the development of recognition as a bridge between capabilities and rights, Ricoeur reminds us that self-recognition implies attestation, which, as we discussed in chapter 5, is expressed in the formula "I believe that I can." He summarized capability as "the power to cause something to happen; it is this power that is liable to self-recognition."[8] Such attestation implies the ascription of an action to an agent: the action Y is ascribed to X, which means X had the power to cause Y. Ricoeur suggests four fundamental capabilities of the self that express this "power to": the power to speak a discourse about things in the world to other people, the power to act and interfere in the state of things in the world, the power to narrate, giving meaning to temporally dispersed actions and events, and finally, the power to impute the moral value of actions to oneself. Although these capabilities might seem restricted to the personal sphere, we should recognize that they necessarily unfold and involve others, which points to a first transition between the personal and institutional levels on the basis of capabilities. A discourse is always addressed to other people. An action typically affects other people and is carried out jointly with and motivated by others. Narratives interweave the actions and sufferings of multiple people as characters in a plot. And the imputation of action involves responsibility toward others: "imputability makes the powerful responsible for the weak and the vulnerable."[9] These fundamental capabilities raise the other as a necessary figure in the process of recognition that confirms and complements "from the outside" what attestation recognizes in the first person.

In the second step of the development of recognition as the fundamental concept connecting capabilities and rights, Ricoeur moves from self-recognition to mutual recognition, emphasizing that the active claim "I strongly believe I can" has as its counterpart the passive form "to be recognized" that expresses a need rather than a claim.[10] And this need to be recognized takes us to the institutional sphere since this "need requires the mediation of institutions providing stability and durability to the process, fulfilling step by step the need to be recognised."[11] The implied other that surfaces in the need for recognition also bridges self-recognition and mutual recognition, as "the category of alterity or otherness assumes the form of reciprocity or mutuality which was lacking (or remained implicit) at the previous stage of self-recognition in terms of capabilities."[12] The mutual recognition of capabilities is based on the fact that capabilities have not only a descriptive dimension but also a normative power that is formalized in legal institutions through the notion of rights.[13] The rights of individuals provide a formal basis for respect for others who participate in the social pact and share the need for mutual recognition. In this sense, rights are the counterpart of mutual recognition in the civic and legal established structures. However, Ricoeur pointed out that these structures often fail to address the situations of injustice and inequity present in the distribution of duties and benefits, which leads to the need for a struggle for recognition in contemporary societies. The struggle for recognition provides one way for establishing mutual recognition of people at an institutional level, particularly in situations of "exclusions due to social inequalities, but also those arising from forms of discrimination inherited from the past that still afflict various minorities."[14] If, on a personal level, self-recognition is the source, through the reflexive movement, of self-esteem, on an institutional level, mutual recognition is the fair distribution of esteem among the institution's participants. The denial of mutual recognition marked by "disdain and humiliation infects the social bond at a level that surpasses rights; this concerns social esteem directed to personal value and to the capacity to pursue happiness in accordance with one's own conception of the good life."[15]

It is precisely the "capacity to pursue happiness in accordance with one's own conception of the good life" that will be the guiding line of our discussion on narrative deliberation in the institutional sphere. Institutions must deliberate on the ways in which digital technologies can maximize individual and collective capacities so that mutual recognition emerges from respectful relationships based on fair and just conditions. Many digital technologies, as we shall discuss, have the potential to level out capacities and increase the possibilities for self-esteem and mutual esteem. However, for Digital Poetics to achieve this ethical goal, narrative deliberation in the institutional sphere needs to shape the development and deployment of emerging technologies based on the ideal of mutual recognition.

The deliberative narrative approach avoids a number of pitfalls in this challenging institutional task. First, it avoids technological solutionism by considering technologies from prospective narratives in which social, economic, and political factors can and should be taken into account. Second, the plurivocality of narrative deliberation is particularly suited to public deliberative environments in which diverse perspectives and expertise need to be integrated. Third, narrative deliberation should always take the idea of a fulfilled life as its starting point and evaluate digital technologies based on this ideal. Finally, the imaginative variations inherent in the exercise of narrative deliberation make it possible to consider and avoid undesirable scenarios and to focus on underprivileged groups that require particular conditions in order to transform digital technologies into generative forces of mutual recognition.

In the next section, we look more closely at the relationship between fostering mutual recognition and "asserting capabilities" based on the conceptual network Ricoeur proposes in *The Course of Recognition*. This theoretical framework will serve as a foundation for our subsequent exploration of the compatibility between the development of new capabilities and digital technologies. The central thesis we intend to explore is that narrative deliberation can fruitfully harness the transformative potential of digital technologies to increase the conditions for mutual recognition in the institutional sphere.

Agency as Power to Act

In *The Course of Recognition*, Ricoeur proposes the concept of recognition as the touchstone for understanding the relationship between personal agency and the structures of just institutions. Good deliberation in the institutional sphere arises from the confluence of two factors: the guarantee of rights for realizing individual capacities and a social organization so that interpersonal and social relationships express mutual esteem among the institution's members. Ricoeur sees the "power to act" as the concept that "binds together" the entire theoretical structure of this intricate network of factors that must be considered for fair deliberation in the institutional sphere. The shift from "power to act" in the individual to the collective sphere affects their respective modes of recognition. In the personal sphere, the "power to act" becomes self-recognition through the reflexive process.[16] In the institutional sphere, the "power to act" is linked to mutual recognition and is based on just social relations. Ricoeur regards the idea of "capabilities," inspired by the work of

Amartya Sen, as the central concept for promoting social justice on the basis of the paradigm of mutual recognition and "power to act."

In particular, Ricoeur considers Sen's relationship between rights and capabilities as a "conceptual revolution," because it contrasts with an evaluation framework based exclusively on utility or well-being. "It is as a real capacity for choice about life that this capability is promoted to the rank of a criterion for evaluating social justice."[17] The key idea lies in the ostensible affirmation of the positivity of the capacity to act, which spills over into the social sphere and creates the basis for mutual recognition. There is a complementarity between the political, economic, and social deliberations that guarantee positive rights and the effective possibility for people to act according to their choices. For Ricoeur, Sen proposes a concept of justice based on capabilities that recognize "the ability to form goals, commitments, values, and so on," which encompasses and extends beyond the exclusive criterion of well-being that other utilitarian strategies of social justice typically propose. There is, therefore, a conceptual harmony between Sen's proposal on an institutional level and Ricoeurian philosophical anthropology, which emphasizes the concept of the capable person.

Ricoeur recalls that Sen's proposal relates to the well-known distinction Isaiah Berlin proposed between "negative freedom" and "positive freedom."[18] Negative freedom refers to freedom from external interference or restrictions. Negative freedom relates to the absence of obstacles or barriers that can prevent individuals from acting according to their own will. In contrast, positive freedom is the freedom to act in such a way as to take control of one's life and fulfill its fundamental purposes. It is not only the absence of restrictions but also the presence of something—a capacity—that allows the individual to act or be in a way that he or she determines. Sen summarizes the relationship between positive freedoms and capabilities as follows: "Positive freedoms are the real freedoms that we have to do things and to be persons, rather than merely being free from interference. They are the freedoms that we require to achieve valuable functionings. Positive freedoms are thus the capabilities of persons, rather than the absence of constraints."[19] Ricoeur clarifies: "Even if this sense of [positive] liberty presupposes the former sense [negative liberty], it adds to it the capacity of a person to lead the life he or she chooses."[20] For our argument, the positive character of capacity is a fundamental idea that links narrative deliberation on the creation of capabilities through digital technologies with the concrete choices that lead people to achieve their ideals of a fulfilled life.

In the context of Digital Poetics, the development of capabilities implies the confluence of fair social conditions, the affordances of digital technologies that facilitate the "power to act," and the free exercise by users of the positive freedoms enhanced by technologies. This necessary confluence of factors confirms the need to think about narrative deliberation in the complementary spheres of deliberation that structure this book. Developers and institutions need to deliberate on the creation of capabilities and the conditions of access and use of capabilities (what Sen calls conversion factors),[21] and users are responsible for freely exercising their positive freedoms by turning capabilities into functionings. This idea creates what Ricoeur called a "rare conjunction"[22] between "collective responsibility," instantiated in our case by public narrative deliberation, and "life choices" that are potentially expanded through new digital technologies.

Sen and Ricoeur are obviously well aware of the risks of the possible subversion of positive freedom into an authoritarian form that pre-determines the actions of individuals and, thereby, removes freedom from their agency and self-determination. Considering this potential misuse of the capabilities approach, Sen emphasizes a distinction between "capabilities" and "functionings."[23] Capabilities are resources that institutions make available for the possibility of positive freedom; they depend on personal choice to incorporate them into their actions and transform them into "functionings." For example, providing an online structure that allows remote health consultations in geographically isolated communities is a capability that can become functioning depending on the choice of each community member.

There are undoubtedly intricate questions about each person's autonomy in choosing to turn capabilities into functioning. For example, institutions need to deliberate carefully on cases involving children and those in vulnerable situations who may lack the necessary conditions for a well-considered choice about whether or not to use capabilities. There is also a more fundamental question about which capabilities should be prioritized and developed.

Sen's proposal leaves room for deliberating about the specific contour of instantiation of the capabilities approach based on the various cultural and political contexts involved.[24] Such delegation to institutional contexts is an essential point of contact between the capabilities approach and narrative deliberation, with the latter offering an appropriate framework for discussing which capabilities should be fostered in a given institutional context.

The importance of critical practical wisdom inherent in our proposal of narrative deliberation is essential for operationalizing the objectives of the capability approach. Sabine Alkire, who works extensively with the capability framework in multidimensional poverty measurement and reduction, writes that "operationalizing is not a one-time thing"[25] but involves a decision-making process that must consider factors such as geographical region, degree of action, and the specific challenges of each socioeconomic context. For Alkire, "many of the residual value judgments in the capability approach will need to be made on the ground repeatedly. They are not of the sort that are susceptible to a magic missing insight."[26] This contention raises important questions about determining, evaluating, balancing, and prioritizing relevant skills in policy/project implementations for which there is no pre-established consensus.[27] This type of contextual decision is the typical form of expression of critical practical wisdom that moves from ideals and principles to the particular issues of each deliberative context.

Martha Nussbaum is another fundamental reference for any theoretical discussion of the capability approach. In *Creating Capabilities: The Human Development Approach*, she develops her own perspective on the capabilities approach, proposes some alternatives or extensions to Sen's perspective, and corroborates several fundamental points of the framework.[28] One of these fundamental points is that the distinction between capabilities and functionings is essential, so that sociopolitical objectives align with the development of capabilities, but people can exercise their freedom to actualize the capabilities available through public institutions in the form of personal functioning.[29]

Among the distinctive features of Nussbaum's proposal is the description of ten central capabilities that she considers fundamental for the approach to be effective. She argues that these ten capabilities are essential if people are to have a fulfilled life that respects the dignity of the human person. They include the ability to live a long and healthy life (life), maintain bodily health and integrity (bodily health and bodily integrity), engage in reasoning and creative expressions (senses, imagination, and thought) and emotional expression (emotions), make informed choices while defining the goals of a flourishing life (practical reason), form social bonds and affiliations (affiliation), appreciate the natural world and other species (other species), enjoy recreational activities and the arts (play), and exercise control over one's environment and life choices (control over one's environment).[30]

Nussbaum stresses that a fulfilled life requires that a minimum threshold be reached for all these capabilities. She leaves room for public deliberation regarding the thresholds that should be achieved. Still, her proposal is much more specific than Sen's, who, as we said, intentionally does not define a particular set of capabilities. Regardless of possible discussions about which capabilities should or should not be central, the multidimensionality of the capabilities approach is more evident in Nussbaum's approach.[31] She invites a holistic analysis of the conditions to be created in public deliberation to foster as many capabilities as possible. She also warns against the danger that creating one capability will undermine other central capabilities. As we will see in the next section, these conditions are important lessons for deliberation in the institutional sphere. When considering how digital technologies can create capabilities for a given community, institutions should consider thresholds, the balance between capabilities, and the possible adverse side effects of developing one capability over others.

Nusbaum also suggests that practical reason and affiliation capabilities are "architectural," because they support, organize, and permeate the other capabilities. This suggestion parallels our proposal for narrative spheres of deliberation about digital technologies. As we saw in chapter 4, applying practical reason is crucial for users to deliberate well on ways of interacting with and integrating the capabilities of digital technologies into their life plans to preserve and increase their ability to act and their self-recognition and self-esteem. In this chapter, we explore how public institutions foster mutual recognition by generating capabilities that involve creating physical or virtual spaces through digital technologies to preserve and promote social bonds and a sense of belonging. To anticipate, a recent example of the way digital technologies can be positive forces in the creation of affiliation was the maintenance of employment and academic relations through videoconferencing during the COVID pandemic.

A final important point of contact between Nussbaum's proposal and narrative deliberation arises in Nussbaum's recourse to narratives to discuss the impact of social initiatives on people's lives. She opens her book with the story of Vasanti, "a small woman [from India] in her early thirties," who used machines to sew saris and, with the support of a nongovernmental organization (NGO) called Self-Employed Women's Organization (SEWA), was able to obtain the resources to establish her financial autonomy, create prospects for her education, and participate in an initiative to combat domestic violence in her community.[32] Vasanti's story offers a prospective narrative model for capacity building, in this particular case involving both a technological artifact and a financial mechanism. On the basis of a prospective narrative model, members of NGOs and government institutions can deliberate on the forms and conditions of applying digital technologies to other narrative contexts.

Sen and Nussbaum's capabilities approach, thus, offers a coherent complement to the Ricoeurian anthropology and ethics that inform our argument about narrative deliberation applied to the institutional sphere of deliberation. The notion of "capabilities" is fundamental in the works of Ricoeur, Sen, and Nussbaum. As we saw in chapter 4, for Ricoeur, capabilities emerge as an inherent aspect of human identity, rooted in the narrative understanding of oneself. For Sen, the development and recognition of capabilities is not just an economic or social undertaking but is intrinsically linked to a philosophical understanding of what it means to be human. According to these perspectives, it is therefore necessary to redefine the metrics of well-being and development, shifting the focus from mere economic growth to the real capabilities people can achieve.

In "Paul Ricoeur and Development Ethics," David Kaplan has recognized and developed several aspects of this affinity and mutual complementarity between Ricoeurian ethics and the capabilities approach.[33] Kaplan suggested that although at first glance Ricoeur and Sen may seem to come from entirely different traditions—Ricoeur with a strong emphasis on hermeneutics and narrative identity and Sen on economic development and human rights—their approaches not only feed off each other but, as we have described, also complement and enrich each other. Both philosophers, in their unique ways, assert that understanding individuals requires recognizing the depth and complexity of their capabilities.

Kaplan also reminds us that Ricoeur's philosophy emphasizes that political imagination, one of the capabilities Nussbaum lists, is fundamental to fostering other capabilities. For Ricoeur, it is through imagination that individuals and societies can envision alternatives, overcome existing deprivations, and forge paths toward new capacities.[34] Such a political imagination is deeply connected to narratives: the stories we tell about ourselves and the stories that shape our communities. As we have argued, Ricoeur's perspective is central to the narrative deliberation approach, which emphasizes the importance of narrative imagination in the decision-making process.

In a study on interpretive anthropology in international development, Ellen Herda corroborates Kaplan's point on the importance of narrative imagination to capabilities development initiative. Herda proposes that the relationship with the members of developing communities should be marked by an effort of narrative imagination that resembles the role that retrospective narrative occupies in narrative deliberation.[35] She contends that narrative imagination, "applied to the development act of working together with the other who faces marginalization, fear, famine, and isolation, provides a medium for both the anthropologist and the local to employ history and fiction into a social reality that brings each to a new place. . . . [T]he relationships created in this place bring narrative to bear on an ethical plane, including remembrance, re-telling, and choosing one's actions."[36] By remembering retrospective narratives and exploring diverse prospective narratives, individuals and communities can challenge the status quo, recognize deprivation, and pave the way for realizing new capabilities through the development and application of digital technologies.

Another essential aspect that Kaplan raises about the contributions of Ricoeur's anthropology to the capabilities approach is the dialectical relationship between fragility and capabilities in the human condition. While capabilities remain at the center of Ricoeur's philosophical exploration through his concept of the "capable human being" (*l'homme capable*), he is also keenly aware of the fragile nature of human existence. Ricoeur directly relates human fragility to the absence of the capacity to act in a way analogous to the concept of deprivation that the capability approach proposes: "Suffering is not defined solely by physical pain, nor even by mental pain, but by the reduction, even the destruction, of the capacity for acting, of being-able-to-act, experienced as a violation of self-integrity."[37] Such suffering, deprivation, and vulnerability should precisely be a focus of social and public initiatives as they seek to mitigate them. It is also essential to recognize that any intervention to create new capabilities can always create or amplify other vulnerabilities, a concern we shall explore further in the following sections. Finally, the inherent dialectic between capabilities and fragilities warns of the risks of fragility of the initiatives to develop capabilities themselves, which are always in danger of becoming mere rhetorical artifice and a perverted political imagination that seeks only the abusive accumulation of power for the interests of a small elite that owns both the platforms and the knowledge created by the new digital technologies.

Against this conceptual backdrop, which brings together Ricoeur's anthropological and ethical philosophy and Sen and Nussbaum's ethics and politics of development, two questions are particularly relevant for further exploring how deliberations in the institutional sphere about digital technologies can promote or disrupt mutual recognition. The first is how to deliberate well about the development and deployment of digital technologies so that they promote new capabilities for communities while preventing existing capabilities from being negatively affected. The second is how to deliberate well to prevent new technologies from becoming ways of increasing the accumulation of power and exacerbating social inequalities.

Both questions reveal how failures in narrative deliberation about the design and application of new technologies weaken mutual recognition. In the first case, scientific, technical, and productive potentials that could be used intentionally to develop technologies aimed at increasing capabilities are often wasted on creating applications aimed solely at innocuous addictive entertainment that feeds a structure of capital accumulation through digital advertising.[38] In this first scenario, flawed narrative deliberation becomes a missed opportunity for digital technological advances to reduce economic, social, and agency inequalities. Such inequalities tear the social fabric and weaken the possibilities for mutual recognition.

In the second case, new technologies positively transform social capabilities, but they are not equally accessible to all community members, creating or widening the gap between those who realize themselves through capabilities fostered by technologies and those excluded from such capabilities. In this case, mutual recognition is interrupted by the gap between economic, social, and political capabilities that is widened by differing access to the economic and intellectual capital required to use digital technologies effectively. Flawed narrative deliberation portends an increase in economic, social, and agency inequities across the digital divide that overlaps with and reinforces other sources of inequity. In the following sections of this chapter, we will discuss these two issues in light of narrative deliberation applied to the institutional sphere.

Developing Capabilities through Digital Technologies

In this section, we investigate digital technologies' potential for generating human capabilities based on the approximation we made in the previous section between the Ricoeurian philosophical anthropology of the capable human being, the capability approach Sen and Nussbaum propose, and the development of mutual recognition as the central objective of Digital Poetics in the institutional sphere. This section's main question is whether and how Digital Poetics is compatible with the development of human capabilities. In doing so, we address the first question from the previous section about deliberating well in the institutional sphere so that digital technologies foster social development and not just the accumulation of capital by the few with technical expertise and computational resources.

In the introduction to *The Capability Approach, Technology and Design*, Ilse Oosterlaken argues that technology and design should play an important role in social justice, human development, and quality of life. She noted that although scholars from various fields have been analyzing these aspects for many years, more recently they have begun to adopt the capability approach to examine them more comprehensively. Oosterlaken specifies that "since roughly 2006, . . . there seems to be a steep increase in the number of scholars using the capability approach for *deliberations* about technology."[39] In the context of this chapter, the use of "deliberation" in the assessment of the application of the capabilities approach to technological issues is particularly auspicious as it underlines the affinity of her analysis with the problem of Digital Poetics.

Furthermore, Oosterlaken points out that much of the discussion about the relationship between technology and capabilities focuses on Information and Communication Technologies (ICTs), which are practically synonymous with the term digital technologies that we have used throughout this book. One of the hypotheses Oosterlaken raises for this predominance is that ICTs are more malleable and can, therefore, be applied to various areas of capacity building, such as health, education, recreation, livelihoods, and democracy. This hypothesis emphasizes the importance of this digital technology characteristic for narrative deliberation in the institutional sphere. It also corroborates our observations in the first chapter about the idiosyncrasies of digital materiality and this materiality's importance to ethical analyses.

Jeroen Van Den Hoven is another theorist who defends the intrinsic relationship between the study of the impact of new technologies on society and the capability approach. He calls this relationship the Capability-Technology Affinity (CAT) thesis.[40] Van den Hoven suggests that one can improve technological and engineering efforts by thoughtfully examining the impact of technology and engineering on human skills and capabilities, as thinkers such as Sen and Nussbaum highlight. Complementarily, one can refine ethical perspectives on human capabilities and functions by looking closely at the conceptualization and implementation of technology. Therefore, the relationship between technology assessment and these ethical perspectives is mutual, and "the afffinity of the CTA thesis is thus bidirectional."[41]

Van den Hoven's arguments bring us back to the central point of this book on Digital Poetics and the inexorable, often neglected, relationship between digital "making" and its ethical aspects, which we investigated extensively in chapter 2. Van den Hoven states that although engineering new technologies and ethical discussions seem to be on separate trajectories, there is a central reason for accelerating efforts to consider them together. Their ultimate goal is (or should be) the same: to promote possibilities for fulfilled lives. Consequently, we should evaluate our successes and contributions in both fields using the same criteria.[42] Referring to Ortega y Gasset's analysis of the teleological relationship between technologies and the ideal of the fulfilled life, Van den Hoven affirms that building the conditions for the fulfilled life should be the terminus point of technical and ethical efforts.[43] Thus, the teleological convergence between ethical deliberation and technological implementation for constructing a shelter against the cold creates the capability for someone to maintain their body temperature at an adequate level. Van den Hoven proposes the concept of agentive amplifiers: "technical artifacts and devices (levers, machines, tools, implements, instruments) . . . create possibilities we would not have without them."[44] Technological artifacts amplify the capacity for human agency in areas as diverse as food, transportation, communication, security, and health. This concept again points to the affinity between the development of technologies and the creation of human capacities.

Van den Hoven offers convincing arguments to support his thesis of the affinity between technologies and the creation of human capabilities. For this affinity to be realized, though, deliberations in the institutional and developer spheres need to build prospective narratives and make decisions that intentionally intertwine aspects of the design, implementation, and deployment of new technologies with the fostering of new capabilities to cultivate fulfilled lives. The affinity between technology and capabilities needs narrative deliberation in the institutional sphere to move from potentiality to effectiveness.

Narrative deliberation must consider the possibility of new technological developments becoming agentive amplifiers that generate new capabilities or conversion factors and enable new functionings for users. Thus, deliberation on the installation for a rural community of a new digital communication channel, such as broadband internet connection, or funding for the development of an artificial intelligence-based assistant for analyzing medical test results must consider prospective narratives that explore the various ways in which such digital technologies will increase or decrease their users' capacity for agency, both individually and collectively.

One salient example of capabilities' development through technologies is the use, in remote regions of Africa, of drones to enhance health-related issues. Traditional methods, which require long and potentially dangerous journeys by road, can be seriously affected by the climate and poor infrastructure. For example, stocks of blood, plasma, and medicine are very sensitive to temperature variations, which are particularly intense in the African region. However, drones offer a fast and reliable alternative, allowing medicines and blood to reach remote areas regardless of the condition of roads or bridges. A company called Zipline designs, manufactures, and operates the world's largest automated delivery system. Using a complex digital system involving artificial intelligence, mobile communications, and drones, Zipline supports efforts such as Rwanda's national blood delivery network and Ghana's COVID vaccine distribution.[45]

The Zips, as the Zipline equipment is called, transport bags of blood from a distribution center in Muhanga, Rwanda, to twenty-one hospitals in the region. In urgent situations, doctors can request blood via the WhatsApp mobile application. Once the request has occurred, the blood is placed in a Zip, which is then launched using a catapult system. For the safe operation of the drone, other digital technology artifacts are involved, such as a mobile app that checks the flying apparatus of the drone using augmented reality. Guided by GPS and coordinated with Rwandan air traffic control, the drone quickly navigates to the designated location. Typically, within an hour of the request, the doctor receives a notification via WhatsApp to collect the package outside the target location. The Zip then safely releases the blood bag using a padded container equipped with a mini parachute. This same equipment and infrastructure was used during the COVID pandemic to speed up contagion testing in a region that is vulnerable to the rapid proliferation of infectious diseases. Zipline seems to be a useful example of an agentive amplifier in the institutional sphere because it increases health capabilities for the population that the service supports. This example also highlights the intertwining of narrative deliberation in the institutional sphere to fund and implement a public system and narrative deliberation in the developers' sphere to design digital technologies to create new health capabilities.

Haenssgen and Ariana argue that digital technologies have both a "generative" and a "transformative" dimension.[46] The generative aspect allows technologies to improve human capabilities directly, while the transformative dimension means that digital technologies impact already existing capabilities, as with Zipline's transformation of local hospital support in Rwanda for blood transfusion. Let's look briefly at how this distinction between the generative and transformative dimensions impacts the capability approach and introduces new elements that narrative deliberation in the institutional sphere must consider.

Within the conceptual framework of the capability approach, inputs are items whose characteristics are associated with the generation of capabilities, so a bicycle, to use Sen's example, brings a more effective capacity for locomotion.[47] Similarly, a vegetable or cereal has the nutritional characteristic that creates the capability of nutritious food. Nonetheless, as we have previously remarked, contextual factors, called conversion factors, filter and transform these inputs. In the case of bicycles, the pathways on which they travel are conversion factors that filter or amplify the generative capacity of that input. Haenssgen and Ariana hypothesize that technologies should be considered both as inputs and conversion factors within a capability development approach.[48]

Taking mobile phones as an example, Haenssgen and Ariana suggest through a qualitative study in rural India and China that the significant role of mobile cellular technology in improving access to healthcare illustrates the profound influence of technology on human development.[49] Access to cell phones allows rural residents to establish other links of affiliation with health service providers and distant family members. However, smartphones also generate or expose other incapabilities that need consideration in narrative deliberation if the benefits of technology are to become capabilities. The generative dimension of smartphones depends on other conversion factors, some technological, others human and social. Haenssgen and Ariana report, for example, the case of an elderly participant in the study who didn't know how to make calls on her cell phone to consult with her medical provider.[50] However, she still valued the new ability to talk on the phone with her daughter, who called her periodically. So, as an input, the smartphone creates the capability for communication and affiliation, but it lacks the conversion factor of digital education so that it can create the capability to access medical services. In addition, mobile devices need other background technologies that operate as conversion factors, such as mobile reception and transmission networks. The study reports, for instance, that neighboring village residents borrowed cell phones because there was no mobile cellular network coverage in their villages, even if they could have their own devices.

Another way of analyzing the impact of smartphones on community capabilities is by looking at how they transform existing capabilities, such as doctor's appointments that previously took place exclusively in person. In this case, smartphones operate as conversion factors or input amplifiers/modulators. As "transformative" artifacts, mobile phones "can facilitate consultations and health advice. This could help overcome spatial marginalization that otherwise limits healthcare access."[51] However, in its capability-transforming dimension, digital technology demands that the deliberative process consider its impact on existing capability structures. In this example of using smartphones as enablers for access to healthcare providers, access to the new digital technology also disrupted healthcare providers who had to adapt their practice to the new demand for mobile device consultations.

The conceptual distinction between the generative and transformative aspects of digital technologies raises a series of practical issues that narrative deliberation needs to address. To explore the transformative impacts of digital technologies that involve complex, multi-layered socio-technological processes, plural deliberation and narrative imagination are required. In addition to its conceptual contributions, the research methodology Haenssgen and Ariana applied exemplifies the potential of narrative deliberation for decisions on adopting new technologies to generate capabilities. For instance, Haenssgen and Ariana's semi-structured interviews resemble a survey of multiple prospective narratives that must be taken into account to imagine a prospective narrative that maximizes the potential of technologies as an "agentive amplifier."[52]

Donovan and Smith studied the deliberative complexities created by the multidimensional impacts of the application of digital technologies in U.K. makerspaces.[53] In particular, they highlight the importance of considering digital technologies within the contexts and social dynamics in which they are made available: "Careful considerations of sociotechnical configurations, conversion factors, and participant diversity are required not only to better distribute expanded capabilities but to mitigate potential harms from those whose capabilities may be diminished."[54] The multidimensional character of the capabilities approach invites a narrative deliberation that explores imaginative variations to take into account not only an obvious benefit of a certain type of capability but also the negative side effects of technologies that diminish the capacity for agency in one or more dimensions, despite the direct intention of increasing agency in another dimension of capability.

In *Technology of the Oppressed: Inequity and the Digital Mundane in Favelas of Brazil*,[55] David Nemer explores how residents of low-income Brazilian favelas use and adapt technologies to combat the challenges they face. Nemer argues that digital technologies become tools for generating capacities and liberating oppressive colonialist structures in ways that counteract socioeconomically disadvantaged regions' lack of agency. Nemer studied places called LAN houses, where favela residents can access computers connected to the internet. This access is similar to cybercafes, which are now obsolete in wealthier centers due to the expansion of broadband internet to all or most homes.

The LAN house scenario is of interest as a case study because a narrative deliberation integrated with the capabilities approach needs to explore how LAN house users' capacity for agency in the favelas would be altered if, for example, broadband access expanded to all residents. Nemer's analysis points to other capabilities the shared space of LAN houses creates that would be negatively affected by simply making broadband available to all households. He reveals, for example, that the spaces of LAN houses have also been essential places for socializing and protection from the violence of the streets dominated by drug cartels. What's more, in LAN houses, residents have found support from other users to learn how to access computer resources, thus reducing the barriers to digital education and the knowledge gap to access the potential brought by digital connections. Thus, a narrative deliberation should create mechanisms in which the broader availability of internet access does not mean the extinction of these spaces where education, security, and social skills flourish.

Collective Deliberation

Advisory Boards

The concept of plural *phronesis* discussed in chapter 3 is also an essential component of narrative deliberation on digital technologies in the institutional sphere. Plural *phronesis* emphasizes that the best decisions are made on the basis of a dialogue between diverse perspectives that inform the decision-making process of applying general principles and rules to particular cases and considers as a criterion the respect due to each person. This approach is especially true for digital technologies that impact multiple facets of human agency and whose understanding requires complementary knowledge of the context in which they are applied, the participants' ideas of a realized life, the institutional structures they will impact, and the technology's technical characteristics. Moreover, because emerging digital technologies generate and transform personal experiences and social structures in often unprecedented and unforeseeable ways at the time legislation and principles are established, the ethical imagination proper to practical wisdom becomes a crucial element in transforming apparent ethical aporias into well-thought-out prospective narratives.

Ricoeur argues that decisions based on practical wisdom are less arbitrary when deliberators consider an inclusive and diverse advisory board of "men and women reputed to be most competent and wisest."[56] Speaking about the complex ethical judgments that judges and doctors have to make in public spheres, Ricoeur claims that they don't have to make these decisions in isolation but should make them, as much as possible, relying on an "advisory board."[57] Although accurate, the English translation of "advisory board" loses direct contact with two fundamental ideas of the original French "*cellule de bon conseil*" that bring important elements to narrative deliberation. The French version highlights the link between this group of deliberators and the teleological idea of a good life in common (*bon*) and the integrative and participatory process that the group has through the notion of a "cell of counsel" (*cellule de conseil*). The plurivocality of the "*cellule de bon conseil*" is a realization of the narrative deliberation that involves the tensional relationship between multiple retrospective and prospective narratives concerning the consequences of the development and application of emerging technologies. Thus, as we have anticipated, deliberation on installing internet access in a remote community must consider not only the technological advances digital connectivity provides but also the possible social and personal impacts of these capabilities. Collective narrative deliberation involving community representatives, local educators, representatives of minorities, representatives of other communities that have gone through the same process, representatives of public authorities, psychologists, and technicians creates a prospective narrative enriched by the diversity of perspectives that should ideally minimize the negative impacts and amplify the positive ones of introducing new technologies into the institutional sphere.

The idea of a "*cellule de bon conseil*" also allows us to address Ricoeur's question related to the diverse traditions represented in plurivocal retrospective narratives. Commenting on this aspect of recent dilemmas in the field of bioethics, which converges with similar problems of new digital technologies, Jean-Philippe Pierron has related this issue to the movement of moral pluralization and secularization of contemporary society: "Ethical pluralism marks the end of the predominance of one moral tradition over others, with the secularization of mores ushering in the coexistence of a plurality of ethical traditions."[58] A fair narrative deliberation must ponder various ethical perspectives in the institutional sphere. Furthermore, this plural retrospective narrative must exist in a dialectical relationship with a prospective narrative that integrates the changes brought about by new narratives. The retrospective and prospective narratives represent the tensional relationship between the poles of traditions and innovations inherent in deliberations in the institutional sphere.

A Productive Tension between Tradition and Innovation

Narrative deliberations in the institutional sphere must take into account the voices and values of the various ethical and political traditions in the context of application. However, these voices that express the forms of meaning transmitted within communities cannot operate as a voice of authority that determines the choice before the fact and closes itself off to any innovation of meaning and capability that can be a source of human development. Traditions are part of the human experience and capture values and priorities concerning the ideal of life in a given sociopolitical context. But they can and should interact with ethical innovations that emerge from the deliberative process to account for significant changes in the existential context of the community, as is the case with emerging digital technologies. As I shall return to, I draw the vocabulary of tradition and innovation in part from Ricoeur.[59]

Emerging technologies create issues requiring innovation to overcome the potential inadequacy of traditional framing of emergent ethical dilemmas. Pierron says, "Ethical discussions, . . . even when they are unprecedented due to the technical innovations that have generated them, cannot disconcert us to the point of preventing us from judging. On the contrary, discussions with our advisory teams reveal the untapped potential that these situations reactivate, inviting us to be inventive and creative."[60] This language perfectly illustrates the intersection between creative imagination and phronetic plurality we propose through narrative deliberation. The different participants bring their perspectives and explore imaginative variations, seeking to create a new framework for giving ethical answers to new questions. Recent Digital Poetics examples are the ethical dilemmas copyright issues of large language models pose or the distribution of responsibility among the various agents of artificial intelligence-based systems.[61]

The dialectic between tradition and innovation also invites us to analyze narrative deliberation in the institutional sphere on the basis of another influential philosophical duality that Ricoeur explores between Jürgen Habermas' ethics of argumentation and the importance that Hans-Georg Gadamer gives to traditions in the hermeneutic process.[62] Ricoeur's criticism of the ethics of argumentation[63] does not lie in its search for the strongest justifications for arguments but in its adopting an exclusively formal approach that separates reasoning from its contextual roots. According to Ricoeur, this abstraction renders the ethics of communication ineffective in navigating the complexity of real-world discourse. He attributes this rigorism to a simplistic interpretation of modernity, which emphasizes a break with an authoritarian past, effectively silencing the productive sources coming from tradition and hindering the development of a more inclusive and nuanced ethical framework for argumentation, one of the central characteristics of narrative deliberation. Nevertheless, one must qualify the term tradition to avoid the possible pathologies linked to the maintenance of power through authoritarian and self-enclosed discourses that have been well identified by Habermas.

To avoid a reductionist analysis of "tradition," Ricoeur pointed out in *Time and Narrative* that "tradition" carries multiple interpretations. One refers to a style of "traditionality" marked by a strong relationship with established practices that nevertheless operates in tension with the possibility of innovation; it represents a dialectic of the established and the innovative. "Traditionality" means "that the temporal distance separating us from the past is not a dead interval but a transmission that is generative of meaning."[64] A second interpretation of tradition refers to "traditions," that is, the multiplicity of customs and practices transmitted within a specific group or community, which can be actively maintained or hardened over time. In this sense, "tradition" is "the things already said, insofar as they are transmitted along the chains of interpretation and reinterpretation."[65] Finally, "Tradition" can be used as a figure of authority that opposes argumentation and critical thinking. Only in this latter sense is the anti-traditionalist crusade of argumentative ethics justified.[66]

For our argument, these distinctions in the interpretation of tradition are fundamental because deliberators in the institutional sphere, although always faced with technological innovation, must consider in their narrative deliberations the diverse traditions and values that create an existing core of meaning for the communities in which the new technologies will potentially be used. Institutional deliberators must apply critical practical wisdom to create decisions that avoid two pathological poles of the spectrum of relationships between significant cultural traditions and technological agentive amplifiers. On one pole of the spectrum, there can be an all-encompassing, out-of-hand rejection of the possibility of productive and positive innovations of meaning based on digital technologies. On the other pole of the spectrum, there can be an exacerbated apology for universalist logical argumentative reasoning that deliberates without considering the cultural traditions and ethical values that structure institutions' meaningful relationships. One of the crucial tasks of narrative deliberation is to configure a prospective narrative that effectively engages the living traditions of the community in the form of a retrospective narrative and integrates them organically with possible technological innovations with the aim of fostering mutual recognition in the institutional sphere.

The Means and Ends Dialectic

Ricoeur was concerned about a mismatch between the human ability to create new technological means and the ability to make sense of the new experiences that technological advances mediate. In "*L'aventure Technique et son Horizon Planétaire*," Ricoeur shares his concern that "the more human beings increase their power, the more acute becomes the problem of ends."[67] I have offered as a proposal to address Ricoeur's concern the focus on mutual recognition and the cultivation of human capacities as a starting point for narrative deliberation on developing and applying new technologies in the institutional sphere. The plasticity of digital technologies allows deliberation to be not just a polarized decision between innovation and tradition but the adjustment of the parameters of design, implementation, and use of technologies in such a way as to respect fundamental values of local traditions and cultural contexts.

Some traditionalist communities, such as the Amish, are an example of deliberation in the public sphere that places traditions and technological innovations in a dialectical relationship within the community and seems to have as its starting point the community's shared ideal of the good life. The distinctiveness of Amish deliberation lies in a communally established process of "testing" new technologies in relation to their impact on community values. Before moving on with this example, it is important to highlight three points to clarify its role in this section's central argument. First, the specific types of choices the Amish make regarding particular digital technologies are not pertinent to our argument, as our point here is only the deliberative process they implement in choosing technological changes. Second, we are not proposing the intrinsic value of Amish technological minimalism. It is reasonable to think that other groups with values different from the Amish deliberate positively about using a vast range of digital technologies, which is not problematic as long as the narrative deliberation considers the dialectic of tradition and innovation and shapes a productive tensional cycle between technological means and ethical ends. Third, in all cases, promoting mutual recognition and developing capabilities remain essential deliberative criteria for Digital Poetics in the institutional sphere. Having made these brief clarifications, let's look at a few points from the Amish deliberative experience regarding incorporating new technologies that provide insights regarding the means-ends dialectic.

The Amish have institutionalized mechanisms to deliberate on technology's impact on their vision of a fulfilled life. Jameson Wetmore[68] has researched the relationship between the Amish and new technologies. He argues that the Amish are not categorically opposed to technology, but rather they question whether technological advances always bring positive results. They don't see technological tools as neutral instruments and understand that the development of technology and the formation of social structures occur simultaneously and significantly impact each other. They find it necessary, therefore, to deliberate on adopting new technologies before they affect personal and community values and identities. To maintain and propagate the values of their community, the Amish have guidelines that dictate which technologies can be used, under what circumstances, how they should be used, and the reasons for their use. Before adopting any new practice or technology, they first assess its alignment with their core beliefs.

Lindsay Ems has also studied the Amish deliberation process on emerging technologies.[69] The Amish's use of smartphones, social networks, and websites surprised Ems. However, she pointed out that "for the Amish, really understanding the functionality and potential social impacts of their tools is essential to deciding whether or not to adopt them."[70] She also explored the dialectic between tradition and ethical innovation in Amish deliberative processes: "Amish discussions about technology adoption . . . represent larger, more complex tensions in communities between holding on to Amish traditions and changing with the times."[71] Wetmore maintains similarly that the Amish "recognize that the technologies they use have become a crucial part of their identity, and they use this link between technology and identity to strengthen their community."[72] The reference to the concept of identity and community operating as criteria for adopting technologies is particularly interesting for our argument. With this example, we point to a group of people who apply a deliberative process that evaluates, based on values chosen as fundamental to a good life, how new technological means impact the social structures and interpersonal relationships of their users.

As we have noted, it is not necessary to establish rules and structures analogous to the Amish or similar communities for a deliberative process on technologies to occur. On the contrary, we want to emphasize that a teleological and deliberative approach that decides on the use of technologies can and should happen in different ways and to varying degrees for all users of digital technologies. The Amish example also allows us to explore the possibility of establishing a more effective relationship between technological innovation and valuing aspects of community traditions that must be preserved to guarantee opportunities for mutual recognition within social groups. The peculiar example of the Amish contrasts with a seemingly predominant view in many Western societies of adopting all digital technologies without the necessary collective deliberation on the impacts of such technologies on the ethical and social values these groups deem relevant ends of the institutional sphere.

However, the diverse and heterogeneous cultural and social contexts of contemporary societies require their own formative and pedagogical concerns that can be substantially more challenging than culturally homogeneous groups. As discussed throughout this book, narrative deliberation provides a framework to address this challenge, but institutions need to create educational mechanisms to prepare their members for productive plural deliberation that aims at respecting the multiplicity of ends within diverse institutions. In "The Task of the Political Educator," Ricoeur talks about the responsibility of educators to prepare new generations for collective deliberation, particularly regarding the tools and techniques of contemporary industrial society. The educator "will no longer simply be a protestor (as he has been and still ought to be) faced with injustice and inequality, old forms of poverty and new forms created by industrialized economics; he ought also to prepare [students] for this responsibility of collective decision."[73]

Collective Capacities

Good narrative deliberation in the institutional sphere must also consider the creation of capacities that all community members share and that become part of their collective identity. In *The Course of Recognition*, Ricoeur speaks of capacities in this expanded sense that "are no longer attested to solely by individuals; they are claimed by collectivities and submitted to public evaluation and approval."[74] In the institutional sphere of deliberation, capacities are not only decided in the community but also transform the community's identity.

The notion of narrative identity applied to communities is productive for thinking about how digital technologies transform not only personal recognition but also mutual and collective recognition. Reflecting on this scope of application of narrative identity, Ricoeur says that "narrative identity is of the greatest importance in an inquiry into the identity of peoples and nations, for it bears the same dramatic and narrative character we all too often confuse with the identity of a substance or a structure."[75] In contrast to an immutable perspective of collective identities, narrative identity integrates the necessary liveliness of events and transformations that affect how communities recognize and value themselves.

The transformations that emerging digital technologies bring about impact collective identities, transforming collective capacities and forms of recognition. A simple example is installing a mechanism to measure drinking water quality for communities lacking sophisticated basic sanitation infrastructure systems. Installing sensors and automated water treatment artifacts changes the history of that community: preventing disease, reducing infant mortality, and affecting collective narrative identity.[76] Similarly, the installation of broadband internet access and teleconsultation devices enables geographically remote communities to access health services more readily and creates new diagnostic and treatment capabilities for all community participants.[77] Through this process of incorporating technologies, the community's narrative identity is altered.

Smart Cities

The term "smart cities" has been used to describe some of these community initiatives to apply digital technologies in an advanced institutional context. According to Halegou,[78] the concept and attributes of smart cities differ widely. However, a key element many share is the emphasis on ICTs (analogous to digital technologies) to gather data about urban areas, create models, and implement autonomous and semi-autonomous processes. Such "smart cities" initiatives aim to enhance, oversee, and optimize city services like transport systems, waste management, energy use, and emergency services. Halegoua and other smart city researchers emphasize the need for collective deliberation that starts from and aims at the good of the community: "If municipal governments want to utilize digital technologies to improve quality of life for all, then technology deployment must be viewed as a means to an end instead of an end itself."[79] By contrast, in practice, big digital technology manufacturers often just impose their products on smart city initiatives, which despite technical merits do not go through a deliberative process that prioritizes the ideal of the good life and the collective identity of communities. The title "smart" is problematic because it can lead to a kind of techno-solutionism that subliminally suggests that the answer to institutional problems lies exclusively in the development and use of more technology. Ben Green warns of the risks of a reductionist vision of "smart" cities: "Technology can be a valuable tool to promote social change, but a technology-driven approach to social progress is doomed from the outset to provide limited benefits or beget unintended negative consequences."[80]

Against this reductionist vision, often linked to the interests of unjust accumulation of power and capital, Digital Poetics in the institutional sphere must deliberate on the impact of new technologies on the collective identity of communities, starting with the development of capabilities that positively impact mutual recognition and the collective identity of communities. Halegoua offers the example of the LocalData application developed jointly by the city of Detroit and three Code for America fellows. LocalData "collects and maps community-generated data about neighborhood conditions such as building and property damage, delivery routes for goods and services, vacancies and abandoned buildings, and environmental issues."[81] This information and models about community fragilities become productive inputs for the deliberation process in the institutional sphere.

Enhanced Capabilities, New Fragilities

In this final section, I briefly point to some areas in which new forms of fragility accompany and emerge from technology-mediated capabilities. The first emerging weakness of digitally mediated capabilities is the possibility of new inequalities created by the application of digital technologies. This problem is linked to the risk of accumulating power through the capabilities that new technologies generate and the pernicious use of this power for domination and exploitation. As Ricoeur suggested in his essay, "The Political Paradox," the risks and diseases of power accumulation transcend historical contingencies, including technical advances: "Techniques change, human relationships evolve depending upon things, and yet power unveils the same paradox, that of a twofold progress in rationality and in possibilities for perversion."[82]

In today's digital technologies era, the changes new technologies bring about are primarily driven by large technology companies that transform their technical power into economic power,[83] thereby influencing the institutional sphere of deliberation, including legislative and regulatory efforts.[84] Once again, we face a paradox of power. On the one hand, the "progress in rationality" that technology companies offer is extremely seductive and convenient and creates new human capabilities. At the same time, this concentration of power generates immense asymmetry of agency and, with it, new personal and social fragilities. In a recent analysis of the relationship between power and technology, Acemoglu and Johnson question the rhetoric that improving technologies entails the corollary of increasing human capabilities. Just as we quoted them to help orient us in chapter 1, it is fitting to quote them again as we approach our close: "A thousand years of history and contemporary evidence make one thing abundantly clear: there is nothing automatic about new technologies bringing widespread prosperity. Whether they do or not is an economic, social, and political choice."[85]

The digital materiality characteristics discussed in the first chapter create new international relations that challenge governmental structures based on physical geography. The physical borders of countries no longer necessarily limit the scope of the impact of technologies. Digital technologies such as social networks, messaging, and videoconferencing applications reach users almost all over the world and influence social and economic systems through digital architectures that permeate the lives of citizens on all continents. Digital materiality also creates new conditions for the accumulation of power from the data derived from the use of digital applications. Although digital technologies provide new functionalities, just to mention a few examples, for communicating, finding service providers, accessing health services, and learning, they create a "behavioral surplus" that captures users' preferences and digital habits.[86] This behavioral surplus is centralized in the databases of giant technology corporations and analyzed by algorithms that can create effective predictive models of users' preferences and institutional processes. Such data accumulation, coupled with the capacity to generate predictive models, intensifies the inequality and asymmetry of agency between communities, users, and large digital technology companies.

The proliferation and improvement of AI-based technologies tend to increase the asymmetry of power between big technology companies and other institutional actors. As discussed in several parts of this book, AI-based artifacts enable a new form of agency where technology companies fully control systems that need fewer and fewer employees to carry out tasks, including the creation of next-generation technologies. Moreover, given the complexity of these types of applications, the technical expertise to handle them is increasingly demanding and centralized in companies with massive research and development budgets. These artificial intelligence systems also require enormous computer processing resources to train the models,[87] which requires substantial capital investments to install and maintain data centers. Without the counterbalance of institutional deliberations that prioritize mutual recognition, the current trajectory of new technology development will lead to a vicious cycle of resource accumulation characterized by the design and commercialization of new technologies with more agency and fewer people involved in the production cycle, which further increases the resources of the holders of technological production capacities.

In addition to the inequality in the ability to create new technologies, a problem also lies in the lack of equity in the conversion factors that allow users to transform the capabilities of new technologies into functionings that generate effective freedoms. The inequality of access to new digital capabilities is evident in global analyses of the availability of the digital infrastructures required for the productive use of new technologies. In 2019, the United Nations World Development Report[88] highlighted that while access to basic digital technologies has increased in developing countries, the digital capabilities gap has widened as the advanced processing, sensing, and communication capabilities of digital technologies remain inaccessible to large segments of the world's population and have become a comparative disadvantage for many. For example, narratives that emphasize the new capabilities that basic mobile services (2G) have brought to sub-Saharan Africa need to be complemented and critiqued by narratives that expose the new vulnerabilities created by the inability to access digital resources that require greater transmission capacity (such as 4G or 5G). These resources are closely linked to core capabilities such as education, agricultural productivity, and online healthcare.

Digital technologies also create inequalities in the labor market. Dissociated from a narrative deliberation aimed at mutual recognition, digital capabilities become exclusive to a self-serving digital elite and create new forms of comparative fragilities in work relations. Acemoglu and Smith emphasized the inequalities and fragilities associated with applying digital technologies in the labor market: "New digital technologies are everywhere and have made vast fortunes for entrepreneurs, executives, and some investors, yet real wages for most workers have scarcely increased."[89]

The negative consequences of an ill-deliberated automation process go beyond the economic impact of using digital agents in the labor market. As Michael Sandel has pointed out,[90] work is an essential dimension of mutual recognition, and the ongoing complexification of jobs takes away from unskilled workers opportunities for attestation and mutual recognition. This line of argumentation leads to the topic of digital automation, which opens up numerous other questions in its own right. A systematic analysis of the disruption of digital technologies in the labor market is beyond the scope of our discussion in this chapter. Nevertheless, this crucial problem reminds us how the characteristics of narrative deliberation and the ethical goal of mutual recognition are essential for the organic development of new capabilities through digital technologies at the institutional level.

To conclude the analysis of the institutional sphere of Digital Poetics, we should return to Ricoeur's discussions on capabilities and rights, where he points to the risk of manipulating capabilities for the accumulation of power: "the exclusion from access to elementary goods is particularly felt as a humiliation generating indignation, anger, and violence."[91] Ricoeur's assessment applies directly to a fallacious rhetoric of human "progress" through digital technologies dissociated from institutional deliberation grounded in mutual recognition, a dissociation that can only widen the valleys of socioeconomic divides with an unfair distribution of capabilities mediated by digital technologies.

The great challenge facing contemporary society involves the development and deployment of digital technologies integrated into complex techno-socio-economic systems that have the potential to promote new human capabilities but can also create new social and individual fragilities due to the reckless use of the power derived from these new technologies. In this historical moment marked by the ubiquity of digital artifacts and the emergence of new synthetic agents based on machine learning algorithms, it is necessary to reconcile technical and ethical aspects through a responsible Digital Poetics based on narrative deliberation that permeates the developers', users', and institutions' spheres.

Notes

1 Paul Ricoeur, *Oneself as Another* (Chicago: University of Chicago Press, 1994), 194.

2 Ricoeur, *Oneself as Another*, 194.

3 See for instance, Martin J. Riedl, Kelsey N. Whipple, and Ryan Wallace, "Antecedents of Support for Social Media Content Moderation and Platform Regulation: The Role of Presumed Effects on Self and Others," *Information, Communication and Society* 25, no. 11 (August 18, 2022): 1632–49. "Blueprint for

an AI Bill of Rights," *The White House*, October 4, 2022, https://www.whitehouse.gov/ostp/ai-bill-of-rights/. Tambiama Madiega, "Artificial Intelligence Act," *European Parliament: European Parliamentary Research Service*, 2021, https://superintelligenz.eu/wp-content/uploads/2023/07/EPRS_BRI2021698792_EN.pdf. Luciano Floridi, "The European Legislation on AI: A Brief Analysis of Its Philosophical Approach," *Philosophy & Technology* 34, no. 2 (June 3, 2021): 215–22.

4 See, for instance, the discussion on the European AI Act, Claudio Novelli et al., "Taking AI Risks Seriously: A New Assessment Model for the AI Act," *AI & Society*, July 12, 2023, https://doi.org/10.1007/s00146-023-01723-z.

5 Andrew D. Selbst, Danah Boyd, Sorelle A. Friedler, Suresh Venkatasubramanian, and Janet Vertesi, "Fairness and Abstraction in Sociotechnical Systems," in *FAT* '19: Proceedings of the Conference on Fairness, Accountability, and Transparency* (New York: Association for Computing Machinery, 2019).

6 Paul Ricoeur, *The Course of Recognition* (Cambridge, MA: Harvard University Press, 2007), 134.

7 Paul Ricoeur, "Capabilities and Rights," in *Transforming Unjust Structures: The Capability Approach*, edited by Mathias Nebel Nicholas Sagovsky Deneulin (Dordrecht: Springer Netherlands, 2006), 18.

8 Ricoeur, "Capabilities and Rights," 18.

9 Paul Ricoeur, "Asserting Personal Capacities and Pleading for Mutual Recognition," in *A Passion for the Possible: Thinking with Paul Ricoeur*, edited by Brian Treanor and Henry Isaac Venema (New York: Fordham University Press, 2010), 23.

10 Ricoeur, "Asserting Personal Capacities and Pleading for Mutual Recognition," 21.

11 Ricoeur, "Asserting Personal Capacities and Pleading for Mutual Recognition," 21.

12 Ricoeur, "Asserting Personal Capacities and Pleading for Mutual Recognition," 21.

13 Ricoeur, "Asserting Personal Capacities and Pleading for Mutual Recognition," 22.

14 Ricoeur, "Asserting Personal Capacities and Pleading for Mutual Recognition," 25.

15 Ricoeur, "Asserting Personal Capacities and Pleading for Mutual Recognition," 25.

16 See the description of the reflexive arc in chapter 5 for a detailed description of this process.

17 Ricoeur, *The Course of Recognition*, 146.

18 Isaiah Berlin, "Two Concepts of Liberty," in *Liberty Reader* (New York: Routledge, 2017), 33–57.

19 Amartya Sen, *Development as Freedom* (Oxford: Oxford University Press, 1999), 75.

20 Ricoeur, *The Course of Recognition*, 143.

21 Amartya Sen, *Inequality Reexamined* (Cambridge, MA: Harvard University Press, 1995), 19–21.

22 Ricoeur, *The Course of Recognition*, 143.

23 Amartya Sen, *Commodities and Capabilities* (Oxford: Oxford University Press, 1999).

24 Ilse Oosterlaken and Jeroen van den Hoven, *The Capability Approach, Technology and Design* (Dordrecht: Springer Science & Business Media, 2012), 3.

25 Sabina Alkire, "Why the Capability Approach?," *Journal of Human Development* 6, no. 1 (March 1, 2005): 127.

26 Alkire, "Why the Capability Approach?" 127.

27 Oosterlaken and Hoven, *The Capability Approach*, 6.

28 Martha Nussbaum, *Creating Capabilities: The Human Development Approach* (Cambridge, MA: Harvard University Press, 2013).

29 Nussbaum, *Creating Capabilities*, 20.

30 Nussbaum, *Creating Capabilities*, 33–34.

31 Nussbaum, *Creating Capabilities*, 32.

32 Nussbaum, *Creating Capabilities*, 2–4.

33 David M. Kaplan, "Paul Ricoeur and Development Ethics," in *A Passion for the Possible: Thinking with Paul Ricoeur*, edited by Brian Treanor and Henry Isaac Venema (New York: Fordham University Press, 2010).

34 Paul Ricoeur, "Reflections on a New Ethos for Europe," *Philosophy & Social Criticism* 21, no. 5–6 (September 1, 1995): 12.

35 Ellen A. Herda, "Narrative Matters Among the Mlabri: Interpretive Anthropology in International Development," in *A Passion for the Possible: Thinking with Paul Ricoeur*, edited by Brian Treanor and Henry Isaac Venema (New York: Fordham University Press, 2010).

36 Herda, "Narrative Matters Among the Mlabri," 129.

37 Ricoeur, *Oneself as Another*, 190.

38 By this, I don't mean to dismiss the value of digital technologies as a form of recreation and often healthy social integration through collaborative video game platforms, for example. After all, "play" is also a central capability for human development. My criticism is directed exclusively at an obtuse application of the potential of digital technologies that uses immense computational resources and developers' creative potential to devise experiences that only aim to maintain

an audience of addicted users that allow the sale of advertising and unlimited concentration of economic value.

39 Oosterlaken and Hoven, *The Capability Approach*, 3 (emphasis added).

40 Jeroen van den Hoven, "Human Capabilities and Technology," in *The Capability Approach, Technology and Design*, edited by Ilse Oosterlaken and Jeroen van den Hoven (Dordrecht: Springer Netherlands, 2012).

41 Van den Hoven, "Human Capabilities and Technology," 29.

42 Van den Hoven, "Human Capabilities and Technology," 23.

43 J. Ortega y Gasset, "Thoughts on Technology," in *Philosophy and Technology: Readings in the Philosophical Problems of Technology*, edited by Carl Mitcham and Robert Mackey (New York: Free Press, 1972).

44 Van den Hoven, "Human Capabilities and Technology," 35.

45 Evan Ackerman and Eliza Strickland, "Medical Delivery Drones Take Flight in East Africa," *IEEE Spectrum* 55, no. 1 (January 2018): 34–35.

46 Marco J. Haenssgen and Proochista Ariana, "The Place of Technology in the Capability Approach," *Oxford Development Studies* 46, no. 1 (January 2, 2018).

47 Amartya Sen, *Resources, Values and Development* (Cambridge, MA: Harvard University Press, 1997), 334.

48 Haenssgen and Ariana. "The Place of Technology in the Capability Approach," 99.

49 Haenssgen and Ariana. "The Place of Technology in the Capability Approach," 105.

50 Haenssgen and Ariana. "The Place of Technology in the Capability Approach," 107.

51 Haenssgen and Ariana. "The Place of Technology in the Capability Approach," 107.

52 Haenssgen and Ariana. "The Place of Technology in the Capability Approach," 105.

53 Cian O'Donovan and Adrian Smith, "Technology and Human Capabilities in UK Makerspaces," *Journal of Human Development and Capabilities* 21, no. 1 (2020): 63–83.

54 O'Donovan and Smith. "Technology and Human Capabilities in UK Makerspaces," 81.

55 David Nemer, *Technology of the Oppressed: Inequity and the Digital Mundane in Favelas of Brazil* (Cambridge, MA: MIT Press, 2022).

56 Ricoeur, *Oneself as Another*, 273.

57 Paul Ricoeur, *Reflections on the Just* (Chicago: University of Chicago Press, 2007), 220.

58 Jean Philippe Pierron, "L'invention Morale et La Sagesse Pratique. Une Lecture de La Petite éthique de Paul Ricoeur," *Études Ricoeuriennes* 10, no. 2 (March 3, 2020): 41. My translation.

59 Ricoeur claims that in fact "tradition is constituted by the interplay of innovation and sedimentation." Tradition is "not the inert transmission of some already dead deposit of material but the living transmission of an innovation always capable of being reactivated by a return to the most creative moments of poetic activity." Paul Ricoeur, *Time and Narrative, Volume 1* (Chicago: University of Chicago Press, 1984), 68.

60 Pierron, "L'invention Morale et La Sagesse Pratique," 43. My translation.

61 Giorgio Franceschelli and Mirco Musolesi, "Copyright in Generative Deep Learning," *Data & Policy* 4 (January 2022): 17.

62 For a synoptic view of the dialectic that Ricoeur proposes between Habermas' critical theory of tradition and Gadamer's hermeneutics, see Paul Ricoeur, "Hermeneutics and the Critique of Ideology," in *Hermeneutics and the Human Sciences: Essays on Language, Action and Interpretation*, edited by John B. Thompson (Cambridge: Cambridge University Press, 1981).

63 Ricoeur, *Oneself as Another*, 333.

64 Ricoeur, *Time and Narrative, Volume 3* (Chicago: University of Chicago Press, 1988), 221. In the first volume of Time and Narrative, Ricoeur also claims that tradition is "not the inert transmission of some already dead deposit of material but the living transmission of an innovation always capable of being reactivated by a return to the most creative moments of poetic activity." Ricoeur, *Time and Narrative Volume 1*, 68.

65 Ricoeur, *Time and Narrative Volume 3*, 221.

66 Ricouer, *Oneself as Another*, 286.

67 Paul Ricoeur, "L'aventure Technique et Son Horizon Planétaire," *Esprit* (2006): 105, https://doi.org/10.3917/espri.0603.0098. My translation.

68 Jameson M. Wetmore, "Amish Technology: Reinforcing Values and Building Community," in *Technology and Society, Second Edition: Building Our Sociotechnical Future*, edited by Deborah G. Johnson and Jameson M. Wetmore (Cambridge, MA: MIT Press, 2021).

69 Lindsay Ems, *Virtually Amish: Preserving Community at the Internet's Margins* (Cambridge, MA: MIT Press, 2022).

70 Ems, *Virtually Amish,* 4.

71 Ems, *Virtually Amish,* 2.

72 Wetmore, "Amish Technology," 250.

73 Paul Ricoeur, "The Tasks of the Political Educator," *Philosophy Today* 17, no. 2 (1973): 148.

74 Ricoeur, *The Course of Recognition*, 134.

75 Paul Ricoeur, *The Just* (Chicago: University of Chicago Press, 2000), 3.

76 Kamarul Hafiz Kamaludin and Widad Ismail, "Water Quality Monitoring with Internet of Things (IoT)," in *2017 IEEE Conference on Systems, Process and Control (ICSPC)* (New York: IEEE, 2017), 18–23.

77 Michael Butzner and Yendelela Cuffee, "Telehealth Interventions and Outcomes Across Rural Communities in the United States: Narrative Review," *Journal of Medical Internet Research* 23, no. 8 (August 26, 2021): e29575.

78 Germaine Halegoua, *Smart Cities* (Cambridge, MA: MIT Press, 2020).

79 Halegoua, *Smart Cities*, xvi.

80 Ben Green, *The Smart Enough City: Putting Technology in Its Place to Reclaim Our Urban Future* (Cambridge, MA: MIT Press, 2019), 4.

81 Halegoua, *Smart Cities*, 71.

82 Paul Ricoeur, "The Political Paradox," in *History and Truth*, edited by Paul Ricœur (Evanston: Northwestern University Press, 1965), 248.

83 Scott Galloway, *The Four: The Hidden DNA of Amazon, Apple, Facebook, and Google* (New York: Penguin, 2018).

84 Emily Birnbaum, "Tech Spent Big on Lobbying Last Year," *Politico*, January 24, 2022, https://www.politico.com/newsletters/morning-tech/2022/01/24/tech-spent-big-on-lobbying-last-year-00001144.

85 Daron Acemoglu and Simon Johnson, *Power and Progress: Our Thousand-Year Struggle Over Technology and Prosperity* (New York: PublicAffairs, 2023), 13.

86 Shoshana Zuboff, *The Age of Surveillance Capitalism: The Fight for a Human Future at the New Frontier of Power* (New York: Profile Books, 2019), chapter 3—The Discovery of Behavioral Surplus, Kindle Edition.

87 Angus Loten, "Rising Data Center Costs Linked to AI Demands," *Wall Street Journal (Eastern Ed.)*, July 13, 2023, https://www.wsj.com/articles/rising-data-center-costs-linked-to-ai-demands-fc6adc0e.

88 Pedro Conceição, *Human Development Report 2019* (New York: United Nations Development Programme, 2019), http://report2019.archive.s3-website-us-east-1.amazonaws.com.

89 Acemoglu and Johnson, *Power and Progress*, 13.

90 M. J. Sandel, *The Tyranny of Merit: What's Become of the Common Good?* (New York: Farrar, Straus and Giroux, 2020).

91 Ricoeur, "Capabilities and Rights," 24.

Bibliography

Access Now. "Human Rights in the Age of Artificial Intelligence." 2018. https://www.accessnow.org/wp-content/uploads/2018/11/AI-and-Human-Rights.pdf.

Acemoglu, Daron and Simon Johnson. *Power and Progress: Our Thousand-Year Struggle Over Technology and Prosperity*. New York: PublicAffairs, 2023.

Ackerman, Evan and Eliza Strickland. "Medical Delivery Drones Take Flight in East Africa." *IEEE Spectrum* 55, no. 1 (January 2018): 34–35.

Alkire, Sabina. "Why the Capability Approach?" *Journal of Human Development* 6, no. 1 (March 1, 2005): 115–35.

Alter, Adam. *Irresistible: The Rise of Addictive Technology and the Business of Keeping Us Hooked*. New York: Penguin, 2017.

Anderson, Janna and Lee Rainie. *The Future of Human Agency*. United States of America: Pew Research Center: Internet, Science & Tech, 2023. https://policycommons.net/artifacts/3456462/the-future-of-human-agency/4256835/.

Anscombe, G. E. M. *Intention*. Cambridge, MA: Harvard University Press, 2000.

Arendt, Hannah. *Between Past and Future: Eight Exercises in Political Thought*. New York: Penguin Books, 1993.

———. *The Human Condition*. 2nd Edition. Chicago: University of Chicago Press, 2013.

Aristotle. *Art of Rhetoric*. Translated by J. H. Freese. Vol. 193. Loeb Classical Library. Cambridge, MA: Harvard University Press, 1926.

———. *Nicomachean Ethics*. Indianapolis: Hackett Publishing, 2019.

———. *Poetics*. Translated by Stephen Halliwell. Vol. XXIII. Loeb Classical Library. Cambridge, MA: Harvard University Press, 1995.

Aschoff, Nicole. *The Smartphone Society: Technology, Power, and Resistance in the New Gilded Age*. Boston: Beacon Press, 2020.

Berlin, Isaiah. "Two Concepts of Liberty." In *Liberty Reader*, 33–57. New York: Routledge, 2017.

Berne, Rosalyn W. and Joachim Schummer. "Teaching Societal and Ethical Implications of Nanotechnology to Engineering Students Through Science Fiction." *Bulletin of Science, Technology & Society* 25, no. 6 (December 1, 2005): 459–68.

Birnbaum, Emily. "Tech Spent Big on Lobbying Last Year." *Politico,* January 24, 2022. https://www.politico.com/newsletters/morning-tech/2022/01/24/tech-spent-big-on-lobbying-last-year-00001144.

Bodle, Jeffrey D., Edward Feldmann, Richard H. Swartz, Zoran Rumboldt, Truman Brown, and Tanya N. Turan. "High-Resolution Magnetic Resonance Imaging: An Emerging Tool for Evaluating Intracranial Arterial Disease." *Stroke*, 2012. https://doi.org/10.1161/STROKEAHA.112.664680.

Boehm, Barry. "A View of 20th and 21st Century Software Engineering." In *Proceedings of the 28th International Conference on Software Engineering*, 12–29. ICSE '06. New York: Association for Computing Machinery, 2006.

Booth, Wayne C. *The Rhetoric of Fiction.* Chicago: University of Chicago Press, 1983.

Bowles, Nellie. "Thermostats, Locks and Lights: Digital Tools of Domestic Abuse." *The New York Times*, June 23, 2018. https://www.nytimes.com/2018/06/23/technology/smart-home-devices-domestic-abuse.html.

Brennan, E. "Parting Ways with the Philosophies of the Subject: Radical Break or Nothing New?" In *Sixth Edition of the Fonds Ricoeur's Summer Workshop*. Paris: Fonds Ricoeur, 2023.

Broussard, Meredith. *More than a Glitch: Confronting Race, Gender, and Ability Bias in Tech.* Cambridge, MA: MIT Press, 2023.

Burton, Emanuelle, Judy Goldsmith, Sven Koenig, Benjamin Kuipers, Nicholas Mattei, and Toby Walsh. "Ethical Considerations in Artificial Intelligence Courses." *AI Magazine*, 2017. https://doi.org/10.1609/aimag.v38i2.2731.

Burton, Emanuelle, Judy Goldsmith, and Nicholas Mattei. "How to Teach Computer Ethics Through Science Fiction." *Communications of the ACM* 61, no. 8 (July 2018): 54–64.

Burton, Emanuelle, Judy Goldsmith, Nicholas Mattei, Cory Siler, and Sara-Jo Swiatek. *Computing and Technology Ethics: Engaging through Science Fiction*. Cambridge, MA: MIT Press, 2023.

Butzner, Michael and Yendelela Cuffee. "Telehealth Interventions and Outcomes Across Rural Communities in the United States: Narrative Review." *Journal of Medical Internet Research* 23, no. 8 (August 26, 2021): e29575.

Calico. "Research & Technology." Accessed October 30, 2023. https://calicolabs.com/research-technology.

Cave, Stephen, Kanta Dihal, and Sarah Dillon. *AI Narratives: A History of Imaginative Thinking about Intelligent Machines*. Oxford: Oxford University Press, 2020.

Chang, Chenliang, Kiseung Bang, Gordon Wetzstein, Byoungho Lee, and Liang Gao. "Toward the Next-Generation VR/AR Optics: A Review of Holographic Near-Eye

Displays from a Human-Centric Perspective." Optica 7, no. 11 (November 20, 2020): 1563–78.

Chown, Eric and Fernando Nascimento. *Meaningful Technologies: How Digital Metaphors Change the Way We Think and Live*. Ann Arbor: Lever Press, 2023.

Coeckelbergh, Mark. *AI Ethics*. Cambridge, MA: MIT Press, 2020.

———. "Language and Technology: Maps, Bridges, and Pathways." *AI & Society* 32, no. 2 (May 1, 2017): 175–89.

Coeckelbergh, Mark and Wessel Reijers. "Cryptocurrencies as Narrative Technologies." *SIGCAS Computers and Society* 45, no. 3 (January 5, 2016): 172–78.

———. "Narrative Technologies: A Philosophical Investigation of the Narrative Capacities of Technologies by Using Ricoeur's Narrative Theory." *Human Studies* 39, no. 3 (September 2016): 325–46.

Conceição, Pedro. *Human Development Report 2019*. New York: United Nations Development Programme, 2019. http://report2019.archive.s3-website-us-east-1.amazonaws.com.

Cuadrado, Félix and Juan C. Dueñas. "Mobile Application Stores: Success Factors, Existing Approaches, and Future Developments." *IEEE Communications Magazine* 50, no. 11 (November 2012): 160–67.

Deneulin, Severine, Mathias Nebel, and Nicholas Sagovsky. *Transforming Unjust Structures: The Capability Approach*. Dordrecht: Springer Science & Business Media, 2006.

Denning, Peter J. and Matti Tedre. *Computational Thinking*. Cambridge, MA: MIT Press, 2019.

DePillis, Lydia and Steve Lohr. "Tinkering With ChatGPT, Workers Wonder: Will This Take My Job?" *The New York Times*, March 28, 2023. https://www.nytimes.com/2023/03/28/business/economy/jobs-ai-artificial-intelligence-chatgpt.html.

Dettinger, Philip, Tobias Kull, Geethika Arekatla, Nouraiz Ahmed, Yang Zhang, Florin Schneiter, Arne Wehling, et al. "Open-Source Personal Pipetting Robots with Live-Cell Incubation and Microscopy Compatibility." *Nature Communications* 13, no. 1 (May 30, 2022): 2999.

Dobson, James E. "Can an Algorithm Be Disturbed? Machine Learning, Intrinsic Criticism, and the Digital Humanities." *College Literature* 42, no. 4 (2015): 543–64.

Domenico, F. "Internet, or the Humanistic Machine." In *The Digital Humanist: A Critical Inquiry*, edited by F. Domenico, N. Teresa, and T. Francesca, 49–84. Santa Barbara: Punctum Books, 2015.

Donner, Jonathan. *After Access: Inclusion, Development, and a More Mobile Internet*. Cambridge, MA: MIT Press, 2015.

Dosse, François. *Paul Ricoeur: Les Sens d'Une Vie.* Paris: La Découverte, 1997.

Else, Holly. "Abstracts Written by ChatGPT Fool Scientists." *Nature* 613, no. 7944 (January 2023): 423.

Ems, Lindsay. *Virtually Amish: Preserving Community at the Internet's Margins.* Cambridge, MA: MIT Press, 2022.

Europea, Comisión. *Communication from the Commission to the European Parliament, the European Council, the Council, the European Economic and Social Committee and the Committee of the Regions Artificial Intelligence for Europe.* European Commission, 2018.

Fiasse, Gaelle. "Aristotle's φρόνησις: A True Grasp of Ends As Well As Means?" *The Review of Metaphysics* 55, no. 2 (2001): 323–37.

Fjeld, Jessica, Nele Achten, Hannah Hilligoss, Adam Nagy, and Madhulika Srikumar. "Principled Artificial Intelligence: Mapping Consensus in Ethical and Rights-Based Approaches to Principles for AI." *Berkman Klein Center Research Publication*, no. 2020-1 (January 15, 2020). https://doi.org/10.2139/ssrn.3518482.

Floridi, Luciano. "The European Legislation on AI: A Brief Analysis of Its Philosophical Approach." *Philosophy & Technology* 34, no. 2 (June 3, 2021): 215–22.

———. *The Fourth Revolution: How the Infosphere Is Reshaping Human Reality.* Oxford: Oxford University Press, 2014.

Ford, Gary. "990 SEI Report on Undergraduate Software Engineering Education." In *Carnegie-Mellon University*. Pittsburgh: Software Engineering Institute, 1990.

Franceschelli, Giorgio and Mirco Musolesi. "Copyright in Generative Deep Learning." *Data & Policy* 4 (January 2022): 17.

Friedman, Batya. "Value-Sensitive Design." *Interactions* 3, no. 6 (December 1, 1996): 16–23.

Friedman, Batya and David G. Hendry. *Value Sensitive Design: Shaping Technology with Moral Imagination.* Cambridge, MA: MIT Press, 2019.

Friedman, Batya and Helen Nissenbaum. "Bias in Computer Systems." *ACM Transactions on Information and System Security* 14, no. 3 (July 1, 1996): 330–47.

Galloway, Scott. *The Four: The Hidden DNA of Amazon, Apple, Facebook, and Google.* New York: Penguin, 2018.

Geraets, T. F. *Rationality to-Day*. Ottawa: University of Ottawa Press, 1979.

Gibney, Elizabeth. "What Google's Winning Go Algorithm Will Do Next." *Nature* 531, no. 7594 (March 17, 2016).

Gomez-Uribe, Carlos A. and Neil Hunt. "The Netflix Recommender System." *ACM Transactions on Management Information Systems*, 2016. https://doi.org/10.1145/2843948.

Gorgoni, Guido. "Reflecting on Identity and Autonomy in a Datafied Society with Paul Ricœur." In *The Challenges of Autonomy and Autonomy as a Challenge Thinking Autonomy in Challenging Time*, edited by Jaroslav Střítecký and Niklas Chimirri, 97–108. Bratislava: Kritika & Kontext, 2022.

Gotterbarn, D. W. "Informatics and Professional Responsibility." *Science and Engineering Ethics* 7, no. 2 (April 2001): 221–30.

———. "Software Engineering Ethics." In *Encyclopedia of Software Engineering*, edited by J. J. Marciniak. New York: Wiley, 2002.

Gotterbarn, D. W., B. Brinkman, C. Flick, and M. S. Kirkpatrick. "ACM Code of Ethics and Professional Conduct Preamble." Accessed December 21, 2023. https://dora.dmu.ac.uk/bitstream/handle/2086/16422/acm-code-of-ethics-and-professional-conduct.pdf?sequence=1.

Green, Ben. *The Smart Enough City: Putting Technology in Its Place to Reclaim Our Urban Future*. Cambridge, MA: MIT Press, 2019.

Greengard, Samuel. *The Internet of Things, Revised and Updated Edition*. Cambridge, MA: MIT Press, 2021.

Haenssgen, Marco J. and Proochista Ariana. "The Place of Technology in the Capability Approach." *Oxford Development Studies* 46, no. 1 (January 2, 2018): 98–112.

Haigh, Thomas and Paul E. Ceruzzi. *The Computer Is Everywhere and Nowhere*. Cambridge, MA: MIT Press, 2021.

Halegoua, Germaine. *Smart Cities*. Cambridge, MA: MIT Press, 2020.

Hall, Crystal, Eric Chown, and Fernando Nascimento. "A Critical, Analytical Framework for the Digital Machine." *Interdisciplinary Science Reviews: ISR* 46, no. 4 (October 2, 2021): 458–76.

Halsema, Annemie. "The Time of the Self: A Feminist Reflection on Ricoeur's Notion of Narrative Identity." *In Time in Feminist Phenomenology*, edited by Christina Schües, Dorothea E. Olkowski, and Helen Fielding, 111–31. Bloomington: Indiana University Press, 2011.

Hamzelou, Jessica. "A Locked-in Man Has Been Able to Communicate in Sentences by Thought Alone." *MIT Technology Review*, March 22, 2022. https://www.technologyreview.com/2022/03/22/1047664/locked-in-patient-bci-communicate-in-sentences/.

Heaven, Will Douglas. "ChatGPT Is Going to Change Education, Not Destroy It." *MIT Technology Review,* April 6, 2023. https://www.technologyreview.com/2023/04/06/1071059/chatgpt-change-not-destroy-education-openai/.

———. "The inside Story of How ChatGPT Was Built from the People Who Made It." *MIT Technology Review*, March 3, 2023. https://www.technologyreview.com/2023/03/03/1069311/inside-story-oral-history-how-chatgpt-built-openai/.

Heikkilä, Melissa. "Responsible AI Has a Burnout Problem." *MIT Technology Review*, October 28, 2022.

Herda, Ellen A. "Narrative Matters Among the Mlabri: Interpretive Anthropology in International Development." In *A Passion for the Possible: Thinking with Paul Ricoeur*, edited by Brian Treanor and Henry Isaac Venema, 129–46. New York: Fordham University Press, 2010.

Honneth, Axel. *The Struggle for Recognition: The Moral Grammar of Social Conflicts.* Cambridge, MA: MIT Press, 1996.

Hoven, Jeroen van den. "Human Capabilities and Technology." In *The Capability Approach, Technology and Design*, edited by Ilse Oosterlaken and Jeroen van den Hoven, 27–36. Dordrecht: Springer Netherlands, 2012.

Hu, Krystal. "ChatGPT Sets Record for Fastest-Growing User Base—Analyst Note." *Reuters,* February 2, 2023. https://www.reuters.com/technology/chatgpt-sets-record-fastest-growing-user-base-analyst-note-2023-02-01/.

Husserl, Edmund. *Cartesian Meditations: An Introduction to Phenomenology.* Netherlands: M. Nijhoff, 2012.

Ihde, Don. "Expanding Hermeneutics." In *Hermeneutics and Science: Proceedings of the First Conference of the International Society for Hermeneutics and Science*, edited by Márta Fehér, Olga Kiss, and László Ropolyi, 345–51. Dordrecht: Springer Netherlands, 1999.

———. *Hermeneutic Phenomenology: The Philosopher of Paul Ricoeur.* Evanston: Northwestern University Press, 1971.

———. *Technology and the Lifeworld: From Garden to Earth*. Bloomington: Indiana University Press, 1990.

Ishiguro, Kazuo. *Klara and the Sun*. London: Faber & Faber, 2022.

Jacobson, Charles, Arie Rip, Thomas J. Misa, and Johan Schot. "Managing Technology in Society: The Approach of Constructive Technology Assessment." *Technology and Culture* 38, no. 3 (July 1997): 813.

Johnson, Deborah G. and Jameson M. Wetmore. *Technology and Society, Second Edition: Building Our Sociotechnical Future*. Cambridge, MA: MIT Press, 2021.

Kamaludin, Kamarul Hafiz and Widad Ismail. "Water Quality Monitoring with Internet of Things (IoT)." In *2017 IEEE Conference on Systems, Process and Control (ICSPC)*, 18–23. New York: IEEE, 2017.

Kaplan, David M. "Paul Ricoeur and Development Ethics." In *A Passion for the Possible: Thinking with Paul Ricoeur,* edited by Brian Treanor and Henry Isaac Venema, 112–28. New York: Fordham University Press, 2010.

———. "Paul Ricoeur and the Philosophy of Technology." *Journal of French and Francophone Philosophy* 16, no. 1/2 (January 26, 2006): 42–56.

Kearney, Richard. "Narrative Imagination: Between Ethics and Poetics." *Philosophy & Social Criticism* 21, no. 5–6 (September 1, 1995): 173–90.

———. "Parsing Narrative—Story, History, Life." *Human Studies* 29, no. 4 (2006): 477–90.

Lake, Brenden. "Finding Solace in Defeat by Artificial Intelligence." *MIT Technology Review*, April 28, 2017. https://www.technologyreview.com/2017/04/28/106009/finding-solace-in-defeat-by-artificial-intelligence/.

Lamptey, Emmanuel and Dorcas Serwaa. "The Use of Zipline Drones Technology for COVID-19 Samples Transportation in Ghana." *HighTech and Innovation Journal* 1, no. 2 (June 1, 2020): 67–71.

Latour, Bruno. "On Technical Mediation." *Common Knowledge* 3, no. 2 (1994): 29–64.

Le Monde. "Le licenciement polémique de Timnit Gebru, qui travaillait chez Google sur les questions d'éthique liées à l'IA." December 7, 2020. https://www.lemonde.fr/pixels/article/2020/12/05/apres-le-licenciement-d-une-chercheuse-noire-et-militante-google-somme-de-s-expliquer_6062328_4408996.html.

Ling, Richard. *New Tech, New Ties: How Mobile Communication Is Reshaping Social Cohesion*. Cambridge, MA: MIT Press, 2010.

Liu, Shubo, Liqing Guo, Heather Webb, Xiao Ya, and Xiao Chang. "Internet of Things Monitoring System of Modern Eco-Agriculture Based on Cloud Computing." *IEEE Access* 7 (2019): 37050–58.

Loten, Angus. "Rising Data Center Costs Linked to AI Demands." *Wall Street Journal (Eastern Ed.)*, July 13, 2023. https://www.wsj.com/articles/rising-data-center-costs-linked-to-ai-demands-fc6adc0e.

Lucivero, Federica, Nina Hallowell, Stephanie Johnson, Barbara Prainsack, Gabrielle Samuel, and Tamar Sharon. "COVID-19 and Contact Tracing Apps: Ethical Challenges for a Social Experiment on a Global Scale." *Journal of Bioethical Inquiry* 17, no. 4 (December 2020): 835–39.

Mac, Ryan, Kellen Browning, and Sheera Frenkel. "The Enduring Afterlife of a Mass Shooting's Livestream Online." *The New York Times*, May 19, 2022. https://www.nytimes.com/2022/05/19/technology/mass-shootings-livestream-online.html.

MacIntyre, Alasdair. *After Virtue*. Notre Dame: University of Notre Dame Press, 1981.

Madiega, Tambiama. "Artificial Intelligence Act." *European Parliament: European Parliamentary Research Service*, 2021. https://superintelligenz.eu/wp-content/uploads/2023/07/EPRS_BRI2021698792_EN.pdf.

Marebane, Senyeki M., Robert T. Hans, Jacqui Coosner, and Ernest Mnkandla. "Perceived Levels of Ethical Responsibilities for a Software Engineer: Computing Academics' Perspective." In *2021 International Conference on Electrical, Computer and Energy Technologies (ICECET)*, 1–6, 2021.

Marsh, Elizabeth, Elvira Perez Vallejos, and Alexa Spence. "The Digital Workplace and Its Dark Side: An Integrative Review." *Computers in Human Behavior* 128 (March 1, 2022): 107118.

McAfee, Andrew and Erik Brynjolfsson. "Opinion." *New York Times*, March 16, 2016. https://www.nytimes.com/2016/03/16/opinion/where-computers-defeat-humans -and-where-the y-cant.html.

McCracken, Harry and Lev Grossman. "Google vs. Death." *Time*. September 30, 2013. https://content.time.com/time/magazine/article/0,9171,2152422,00.html.

Metz, Cade and Daisuke Wakabayashi. "Google Researcher Says She Was Fired Over Paper Highlighting Bias in A.I." *The New York Times,* Deccember 3, 2020. https:// www.nytimes.com/2020/12/03/technology/google-researcher-timnit-gebru.html.

Mohanty, Jitendra Nath. "Method of Imaginative Variation in Phenomenology." *Thought Experiments in Science and Philosophy*, 1991, 261–72.

Moustakas, Clark. *Phenomenological Research Methods*. London and New Delhi: SAGE Publications, 1994.

Nascimento, Fernando and Anya Workman. "Rebalancing the Digital Convenience Equation through Narrative Imagination." *International Conference on Computer Ethics* 1, no. 1 (2023). https://soremo.library.iit.edu/index.php/CEPE2023/article/ view/271.

Nemer, David. *Technology of the Oppressed: Inequity and the Digital Mundane in Favelas of Brazil.* Cambridge, MA: MIT Press, 2022.

Nissenbaum, Helen. "Protecting Privacy in an Information Age: The Problem of Privacy in Public." *Law and Philosophy* 17, no. 5/6 (November 1998): 559.

Nussbaum, Martha C. *Creating Capabilities: The Human Development Approach*. Cambridge, MA: Harvard University Press, 2013.

———. *The New Religious Intolerance.* Cambridge, MA: Harvard University Press, 2012.

O'Donovan, Cian and Adrian Smith. "Technology and Human Capabilities in UK Makerspaces." *Journal of Human Development and Capabilities* 21, no. 1 (2020): 63–83.

O'Mara, Margaret. *The Code: Silicon Valley and the Remaking of America*. New York: Penguin, 2020.

O'Neil, Cathy. *Weapons of Math Destruction: How Big Data Increases Inequality and Threatens Democracy*. New York: Crown, 2016.

Ortega y Gasset, J. "Thoughts on Technology." In *Philosophy and Technology: Readings in the Philosophical Problems of Technology*, edited by Carl Mitcham and Robert Mackey. New York: Free Press, 1972.

Pakaluk, Michael, ed. "Thinking-Related Virtue (Nicomachean Ethics, Book 6)." In *Aristotle's Nicomachean Ethics: An Introduction*, 206–32. Cambridge Introductions to Key Philosophical Texts. Cambridge: Cambridge University Press, 2005.

Pandav, Krunal, Austen G. Te, Nir Tomer, Sujit S. Nair, and Ashutosh K. Tewari. "Leveraging 5G Technology for Robotic Surgery and Cancer Care." *Cancer Reports* 5, no. 8 (August 2022): e1595.

Pariser, Eli. *The Filter Bubble: What the Internet Is Hiding from You*. New York: Penguin Books, 2012.

Pereira, Roberto and Maria Cecília Calani Baranauskas. "A Value-Oriented and Culturally Informed Approach to the Design of Interactive Systems." *International Journal of Human-Computer Studies* 80 (August 1, 2015): 66–82.

Picardie, Université de, Jeffrey Andrew Barash, and Mireille Delbraccio. *La Sagesse Pratique: Autour de L'oeuvre de Paul Ricoeur: Colloque International, Université de Picardie Jules Verne, Amiens, 5-7 Mars 1997*. Centre national de documentation pédagogique, 1998.

Pierron, Jean-Philippe. "L'invention Morale et La Sagesse Pratique. Une Lecture de La Petite éthique de Paul Ricoeur." *Études Ricoeuriennes* 10, no. 2 (March 3, 2020): 36–51.

Plato. *The Statesman*. Cambridge, MA: Cambridge University Press, 1995.

Pressman, Roger S. and Bruce R. Maxim. *Software Engineering: A Practitioner's Approach*. New York: McGraw-Hill Education, 2019.

Priestley, Mark. *A Science of Operations: Machines, Logic and the Invention of Programming*. London: Springer Science & Business Media, 2011.

Ratcliffe, Susan. "Steve Jobs." In *Oxford Essential Quotations*. Oxford University Press, 2016. https://www.oxfordreference.com/view/10.1093/acref/9780191826719.001.0001/q-oro-ed4-00005922.

Reeve, C. D. C. "Aristotle on the Virtues of Thought." In *The Blackwell Guide to Aristotle's Nicomachean Ethics*, edited by Richard Kraut, 199–218. New York: John Wiley & Sons, 2008.

Regalado, Antonio. "Saudi Arabia Plans to Spend $1 Billion a Year Discovering Treatments to Slow Aging." *MIT Technology Review*, June 7, 2022. https://www.technologyreview.com/2022/06/07/1053132/saudi-arabia-slow-aging-metformin/.

Reijers, Wessel and Mark Coeckelbergh. *Narrative and Technology Ethics*. New York: Palgrave Macmillan Cham, 2020.

Reijers, Wessel, David Wright, Philip Brey, Karsten Weber, Rowena Rodrigues, Declan O'Sullivan, and Bert Gordijn. "Methods for Practising Ethics in Research and Innovation: A Literature Review, Critical Analysis and Recommendations." *Science and Engineering Ethics* 24, no. 5 (October 2018): 1437–81.

Ricoeur, Paul. "À La Gloire de la Phronesis (Éthique a Nicomaque, Livre VI)." In *La vérité pratique: Aristote, Ethique à Nicomaque, Livre VI*, edited by Jean-Yves Chateau, 13–22. Paris: Vrin, 1997.

———. "Architecture and Narrativity." Translated by E. Brennan, R. Carney, and S. Lelièvre *Études Ricoeuriennes/Ricoeur Studies* 7, no. 2 (2016): 20–30.

———. "Architecture et Narrativité." *Études Ricoeuriennes* 7, no. 2 (February 1, 2017): 20–30.

———. "Asserting Personal Capacities and Pleading for Mutual Recognition." In *A Passion for the Possible: Thinking with Paul Ricoeur*, edited by Brian Treanor and Henry Isaac Venema. New York: Fordham University Press, 2010.

———. "Capabilities and Rights." In *Transforming Unjust Structures: The Capability Approach*, edited by Mathias Nebel Nicholas Sagovsky Deneulin, 17–26. Dordrecht: Springer Netherlands, 2006.

———. "Éthique et Morale." *Revista Portuguesa de Filosofia* 46, no. 1 (1990): 5–17.

———. *Fallible Man*. New York: Fordham University Press, 1965.

———. *Figuring the Sacred: Religion, Narrative, and Imagination*. Minneapolis: Fortress Press, 1995.

———. *From Text to Action: Essays in Hermeneutics, II*. Vol. 2. London: Bloomsbury Publishing, 2008.

———. "Hermeneutics and the Critique of Ideology." In *Hermeneutics and the Human Sciences: Essays on Language, Action and Interpretation*, edited by John B. Thompson, 63–100. Cambridge: Cambridge University Press, 1981.

———. *Hermeneutics and the Human Sciences*. Cambridge: Cambridge University Press, 2016.

———. "Hermeneutics and the World of the Text." In *Hermeneutics: Writings and Lectures—Volume 2*, translated by David Pellauer. Cambridge: Polity, 2013.

———. "Intellectual Autobiography." In *The Philosophy of Paul Ricoeur*, edited by Lewis Edwin Hahn. Chicago: Open Court, 1995.

———. "La Raison Pratique." In *Rationality To-Day*, edited by T. F. Geraets, 225–48. Ottawa: University of Ottawa Press, 1979.

———. "L'aventure Technique et Son Horizon Planétaire." *Esprit*, 2006. https://doi.org/10.3917/espri.0603.0098.

———. "Life in Quest of Narrative." In *On Paul Ricoeur. Narrative and Interpretation.*, edited by David Wood, 20–33. London and New York: Routledge, 1991.
———. *Memory, History, Forgetting.* Chicago: University of Chicago Press, 2004.
———. "Narrative Identity." *Philosophy Today* 35, no. 1 (1991): 73–81
———. "Narrative Time." *Critical Inquiry* 7, no. 1 (October 1, 1980): 169–90.
———. *Oneself as Another.* Chicago: University of Chicago Press, 1994.
———. *Political and Social Essays.* Athens: Ohio University Press, 1975.
———. "Reflections on a New Ethos for Europe." *Philosophy & Social Criticism* 21, no. 5–6 (September 1, 1995): 3–13.
———. *Reflections on the Just.* Chicago: University of Chicago Press, 2007.
———. *Réflexion Faite: Autobiographie Intellectuelle.* Paris: Éditions Esprit, 1995.
———. *The Conflict of Interpretations: Essays in Hermeneutics.* Evanston: Northwestern University Press, 1974.
———. *The Course of Recognition.* Cambridge, MA: Harvard University Press, 2007.
———. *The Just.* Chicago: University of Chicago Press, 2000.
———. "The Memory of Suffering." *Criterion* 28, no. 2 (1989): 4.
———. "The Model of the Text: Meaningful Action Considered as a Text." *New Literary History* 5, no. 1 (1973): 91–117.
———. "The Political Paradox." In *History and Truth*, edited by Paul Ricoeur. Evanston: Northwestern University Press, 1965.
———. *The Rule of Metaphor: Multi-Disciplinary Studies of the Creation of Meaning in Language.* Toronto: University of Toronto Press, 1978.
———. "The Tasks of the Political Educator." *Philosophy Today* 17, no. 2 (1973): 142–52.
———. "The Text as Dynamic Identity." *Identity of the Literary Text*, n.d.
———. *Time and Narrative, Volume 1.* Chicago: University of Chicago Press, 1984.
———. *Time and Narrative, Volume 3.* Chicago: University of Chicago Press, 1988.
Ricœur, Paul, François Azouvi, and Marc B. de Launay. *Critique and Conviction: Conversations with François Azouvi and Marc de Launay*. New York: Columbia University Press, 1998.
Riedl, Martin J., Kelsey N. Whipple, and Ryan Wallace. "Antecedents of Support for Social Media Content Moderation and Platform Regulation: The Role of Presumed Effects on Self and Others." *Information, Communication and Society* 25, no. 11 (August 18, 2022): 1632–49.
Romele, Alberto. *Digital Hermeneutics: Philosophical Investigations in New Media and Technologies.* New York: Routledge, 2019.

———. "Narrative Identity and Social Networking Sites." *Études Ricoeuriennes* 4, no. 2 (January 3, 2014): 108–22.

Saad, Lydia. "Americans Have Close but Wary Bond with Their Smartphone." *Gallup*, June 20, 2022. https://news.gallup.com/poll/393785/americans-close-wary-bond-smartphone.aspx.

Sandel, M. J. *The Tyranny of Merit: What's Become of the Common Good?* New York: Farrar, Straus and Giroux, 2020.

Schatzberg, Eric. *Technology: Critical History of a Concept.* Chicago: University of Chicago Press, 2018.

Schön, Donald A. "Generative Metaphor: A Perspective on Problem-setting in Social Policy." In *Metaphor and Thought*, edited by Andrew Ortony, 137–63. Cambridge: Cambridge University Press, 1993. doi:10.1017/CBO9781139173865.011.

Schrage, Michael. *Recommendation Engines.* Cambridge, MA: MIT Press, 2020.

Selbst, Andrew D., Danah Boyd, Sorelle A. Friedler, Suresh Venkatasubramanian, and Janet Vertesi. "Fairness and Abstraction in Sociotechnical Systems." In *FAT* '19: Proceedings of the Conference on Fairness, Accountability, and Transparency*, 59–68. New York: Association for Computing Machinery, 2019.

Sen, Amartya. *Commodities and Capabilities*. Oxford: Oxford University Press, 1999.

———. *Development as Freedom.* Oxford: Oxford University Press, 1999.

———. Inequality Reexamined. Cambridge, MA: Harvard University Press, 1995.

———. *Resources, Values and Development*. Cambridge, MA: Harvard University Press, 1997.

Skirpan, Michael, Nathan Beard, Srinjita Bhaduri, Casey Fiesler, and Tom Yeh. "Ethics Education in Context: A Case Study of Novel Ethics Activities for the CS Classroom." In *Proceedings of the 49th ACM Technical Symposium on Computer Science Education*, 940–45. SIGCSE '18. New York: Association for Computing Machinery, 2018.

Smith, Brent and Greg Linden. "Two Decades of Recommender Systems at Amazon.com." *IEEE Internet Computing* 21, no. 3 (May 2017): 12–18.

Stapleton, Larry. "Ethical Decision Making in Technology Development: A Case Study of Participation in a Large-Scale Information Systems Development Project." *AI & Society* 22, no. 3 (January 2008): 405–29.

Stiegler, Christian. *The 360° Gaze: Immersions in Media, Society, and Culture.* Cambridge, MA: MIT Press, 2021.

Taylor, George H. "Prospective Political Identity." In *Paul Ricoeur in the Age of Hermeneutical Reason*, edited by Roger W. H. Savage, 123–37. Lanham: Lexington Books, 2015.

Taylor, George H. and Zeta Books. "Delineating Ricoeur's Concept of Utopia." *Social Imaginaries* 3, no. 1 (2017): 41–60.

The White House. "Blueprint for an AI Bill of Rights." October 4, 2022. https://www.whitehouse.gov/ostp/ai-bill-of-rights/.

Thompson, Neil C., Shuning Ge, and Gabriel F. Manso. "The Importance of (Exponentially More) Computing Power," *arXiv [cs.AR]*, June 28, 2022. https://doi.org/10.48550/arXiv.2206.14007.

Tiku, Nitasha. "Google Hired Timnit Gebru to be an Outspoken Critic of Unethical AI. Then She Was Fired for it." *The Washington Post*, December 23, 2020. https://www.washingtonpost.com/technology/2020/12/23/google-timnit-gebru-ai-ethics/.

Turkle, Sherry. *Alone Together: Why We Expect More from Technology and Less from Each Other*. New York: Basic Books, 2017.

Van Den Boomen, Marianne. *Transcoding the Digital: How Metaphors Matter in New Media.* Amsterdam: Institute of Network Cultures, 2014.

Verbeek, Peter-Paul. "Don Ihde: The Technological Lifeworld." In *American Philosophy of Technology: The Empirical Turn*, edited by Hans Achterhuis, translated by Robert P. Crease, 119–46. Bloomington: Indiana University Press, 2001.

———. "Materializing Morality: Design Ethics and Technological Mediation." *Science, Technology & Human Values* 31, no. 3 (May 1, 2006): 361–80.

———. "Morality in Design: Design Ethics and the Morality of Technological Artifacts." In *Philosophy and Design: From Engineering to Architecture*, edited by Peter Kroes, Pieter E. Vermaas, Andrew Light, and Steven A. Moore, 91–103. Dordrecht: Springer Netherlands, 2008.

———. "Toward a Theory of Technological Mediation." *Technoscience and Postphenomenology: The Manhattan Papers 189* (2015). https://books.google.com/books?hl=en&lr=&id=oJcpCwAAQBAJ&oi=fnd&pg=PA189&dq=peter-paul+verbeek&ots=rhcfUPMSzY&sig=nLUduF5MrqsOaXWb3ib1j9rlb1I.

———. *What Things Do: Philosophical Reflections on Technology, Agency, and Design*. University Park: Penn State Press, 2010.

Villani, Cédric, Yann Bonnet, Marc Schoenauer, Charly Berthet, Francois Levin, Anne Charlotte Cornut, and Bertrand Rondepierre. *For a Meaningful Artificial Intelligence: Towards a French and European Strategy*. France: Conseil national du numérique, 2018.

Wall, John. "Phronesis as Poetic: Moral Creativity in Contemporary Aristotelianism." *The Review of Metaphysics* 59, no. 2 (2005): 313–31.

———. "Phronesis, Poetics, and Moral Creativity." *Ethical Theory and Moral Practice: An International Forum* 6, no. 3 (September 1, 2003): 317–41.

Weizenbaum, J. "On the Impact of the Computer on Society." *Science* 176, no. 4035 (May 12, 1972): 609–14.

Wessel, Max and Nicole Helmer. "A Crisis of Ethics in Technology Innovation." *The Next Age of Disruption*, 2021. https://doi.org/10.7551/mitpress/13768.003.0008.

Wetmore, Jameson M. "Amish Technology: Reinforcing Values and Building Community." In *Technology and Society, Second Edition: Building Our Sociotechnical Future*, edited by Deborah G. Johnson and Jameson M. Wetmore. Cambridge, MA: MIT Press, 2021.

Wolff, Ernst. *Between Daily Routine and Violent Protest: Interpreting the Technicity of Action.* Berlin and Boston: De Gruyter, 2021.

———. "Ricoeur's Polysemy of Technology and Its Reception." In *Interpreting Technology: Ricoeur on Questions Concerning Ethics and Philosophy of Technology*, edited by Wessel Reijers, Alberto Romele, and Mark Coeckelbergh. Lanham: Rowman & Littlefield, 2021.

———. "'Technology' as the Critical Social Theory of Human Technicity." *Journal of Philosophical Research* 41 (October 6, 2016): 333–69.

Wong, Julia Carrie. "More than 1,200 Google Workers Condemn Firing of AI Scientist Timnit Gebru." *The Guardian,* December 7 2020. https://www.theguardian.com/technology/2020/dec/04/timnit-gebru-google-ai-fired-diversity-ethics.

Wright, Georg Henrik von. *Explanation and Understanding.* Ithaca: Cornell University Press, 2004.

Zuboff, Shoshana. *The Age of Surveillance Capitalism: The Fight for a Human Future at the New Frontier of Power.* New York: PublicAffairs, 2019.

Index

About the Author

Fernando Nascimento is an assistant professor of digital and computational studies at Bowdoin College, researching the intersections of digital technologies and ethics. His current research focuses primarily on how digital technologies can enhance human capabilities. He also directs the Digital Ricoeur project, which applies digital text analysis to support worldwide academic research on the works of the French philosopher Paul Ricoeur.

www.ingramcontent.com/pod-product-compliance
Ingram Content Group UK Ltd.
Pitfield, Milton Keynes, MK11 3LW, UK
UKHW020215290926
15470UKWH00001B/192

9 781666 942231